ENGINEERING MAINTENANCE

A Modern Approach

ENGINEERING MAINTENANCE

A Modern Approach

B.S. Dhillon, Ph.D.

CRC PRESS

Boca Raton   London   New York   Washington, D.C.

Library of Congress Cataloging-in-Publication Data

Dhillon, B.S.
    Engineering maintenance : a modern approach / by B.S. Dhillon.
        p. cm.
    Includes bibliographical references and index.
    ISBN 1-58716-142-7
    1. Plant maintenance—Management. I. Title.

TS192 .D47 2002
658.2′02—dc21   2001052634

This book contains information obtained from authentic and highly regarded sources. Reprinted material is quoted with permission, and sources are indicated. A wide variety of references are listed. Reasonable efforts have been made to publish reliable data and information, but the authors and the publisher cannot assume responsibility for the validity of all materials or for the consequences of their use.

Visit the CRC Press Web site at www.crcpress.com

© 2002 by CRC Press LLC

No claim to original U.S. Government works
International Standard Book Number 1-58716-142-7
Library of Congress Card Number 2001052634
Printed in the United States of America   2 3 4 5 6 7 8 9 0
Printed on acid-free paper

Preface

Engineering maintenance is an important sector of the economy. Each year U.S. industry spends well over $300 billion on plant maintenance and operation, and in 1997 the U.S. Department of Defense's budget request alone included $79 billion for operation and maintenance. Furthermore, it is estimated that approximately 80% of the industry dollars is spent to correct chronic failures of machines, systems, and people. The elimination of many of these chronic failures through effective maintenance can reduce the cost between 40 and 60%.

This century will usher in a broader need for equipment management—a cradle-to-grave strategy to preserve equipment functions, avoid the consequences of failure, and ensure the productive capacity of equipment. This cannot be achieved by simply following the traditional approach to maintenance effectively—human error in maintenance, quality and safety in maintenance, software maintenance, reliability-centered maintenance, maintenance costing, reliability, and maintainability also must be considered.

Today, a large number of books are available on maintenance, but to the best of my knowledge, none covers all the areas listed above. Material on such topics is available either in technical articles or in specialized books, but not in a single volume. In order to perform the maintenance function effectively, knowledge of these topics is essential, but maintenance professionals find it difficult to obtain such information in a single maintenance text.

The main objective of this book is to cover all the above and other related current topics in a single volume in addition to the traditional topics of engineering maintenance. The book focuses on the structure of concepts rather than the minute details. The sources of most of the material are given in references, which will be useful to readers who desire to delve deeper into specific areas.

Chapter 1 presents various introductory aspects concerning engineering maintenance including engineering objectives, engineering maintenance in the 21st century, and maintenance-related facts and figures. Chapter 2 reviews the basic probability theory and other pertinent mathematical topics that will help the reader understand subsequent chapters of the book. Chapter 3 discusses various aspects related to maintenance management and control, including department functions and organizations, elements of effective management, management control indices, and project control methods.

Chapter 4 is devoted to preventive maintenance (PM) and covers topics such as preventive maintenance elements; steps for establishing a PM program; and PM measures, models, and advantages and disadvantages. Chapter 5 presents various aspects of corrective maintenance (CM) ranging from CM types and measures to CM mathematical models. Chapter 6 is devoted to the important subject of reliability centered maintenance (RCM) and covers topics such as RCM goals and principles, RCM process, RCM components, and RCM program effectiveness indicators.

Inventory control in maintenance is presented in Chapter 7. This chapter covers topics such as inventory types and purposes, inventory control models, safety stock, and estimation of spare part quantity. Chapter 8 and 9 are devoted to human error in maintenance and quality and safety in maintenance, respectively. Some of the topics covered in Chapter 8 are facts and figures on human error in maintenance, maintenance error in system life cycle, guidelines for reducing human error, and techniques for predicting the occurrence of human error. Chapter 9 includes topics such as the need for quality maintenance processes, maintenance work quality, quality control charts for use in maintenance, post maintenance testing, safety and maintenance tasks, guidelines for equipment designers to improve safety in maintenance, and maintenance personnel safety.

Chapter 10 presents various aspects concerning maintenance costing, including reasons for maintenance costing, factors influencing cost, labor and material cost estimation, cost estimation models, and cost data collection. Chapter 11 presents an important area of modern maintenance, i.e., software maintenance. Some of the topics relating to software maintenance are types of software maintenance, software maintenance problems, software maintenance tools and techniques, and software maintenance costing.

Chapters 12 and 13 are devoted to two areas closely related to maintenance, i.e., reliability and maintainability. Chapter 12 covers reliability measures, reliability networks, and reliability analysis methods. Chapter 13 includes maintainability management in system life cycle, maintainability design characteristics, maintainability measures and functions, and common errors related to maintainability design.

This book will be useful to senior level undergraduate and graduate students in mechanical and industrial engineering; maintenance and operations, engineers; college and university level teachers; students and instructors of short courses in engineering maintenance; and equipment designers, managers, manufacturers, and users.

The author is deeply indebted to many friends, colleagues, and students for their interest and encouragement throughout this project. I thank my children, Jasmine and Mark, for their patience and intermittent disturbances leading to desirable coffee and other breaks. And last, but not least, I thank my wife, Rosy, for typing various portions of this book, editorial input, proofreading, and tolerance.

B.S. Dhillon

Dedication

---

*This book is affectionately dedicated to my wife, **Rosy***

Table of Contents

Chapter 6: Reliability Centered Maintenance

Chapter 7: Inventory Control in Maintenance

Chapter 10: Maintenance Costing

Chapter 11: Software Maintenance

1 Introduction

BACKGROUND

Since the Industrial Revolution, maintenance of engineering equipment in the field has been a challenge. Although impressive progress has been made in maintaining equipment in the field in an effective manner, maintenance of equipment is still a challenge due to factors such as size, cost, complexity, and competition. Needless to say, today's maintenance practices are market driven, in particular for the manufacturing and process industry, service suppliers, and so on.[1] An event may present an immediate environmental, performance, or safety implication. Thus, there is a definite need for effective asset management and maintenance practices that will positively influence critical success factors such as safety, product quality, speed of innovation, price, profitability, and reliable delivery.

Each year billions of dollars are spent on equipment maintenance around the world. Over the years, many new developments have taken place in this area. The terms "maintenance" and "maintenance engineering" may mean different things to different people. For example, the U.S. Department of Defense sees maintenance engineering as a discipline that assists in acquisition of resources needed for maintenance, and provides policies and plans for the use of resources in performing or accomplishing maintenance.[2] In contrast, maintenance activities are viewed as those that use resources in physically performing those actions and tasks attendant on the equipment maintenance function for test, servicing, repair, calibration, overhaul, modification, and so on.

Comprehensive lists of publications on maintenance and maintenance engineering are given in References 3 and 4.

MAINTENANCE AND MAINTENANCE ENGINEERING OBJECTIVES

Even though maintenance engineering and maintenance have the same end objective or goal (i.e., mission-ready equipment/item at minimum cost), the environments under which they operate differ significantly. More specifically, maintenance engineering is an analytical function as well as it is deliberate and methodical. In contrast, maintenance is a function that must be performed under normally adverse circumstances and stress, and its main objective is to rapidly restore the equipment to its operational readiness state using available resources. Nonetheless, the contributing objectives of maintenance engineering include: improve maintenance operations, reduce the amount and frequency of maintenance, reduce the effect of complexity, reduce the maintenance skills required, reduce the amount of supply support, establish optimum frequency

1

and extent of preventive maintenance to be carried out, improve and ensure maximum utilization of maintenance facilities, and improve the maintenance organization.[2]

This book directly or indirectly covers both maintenance and maintenance engineering and their objectives.

MAINTENANCE FACTS AND FIGURES

Some the important facts and figures directly or indirectly associated with engineering maintenance are as follows:

- Each year over $300 billion are spent on plant maintenance and operations by U.S. industry, and it is estimated that approximately 80% of this is spent to correct the chronic failure of machines, systems, and people.[5]
- In 1970, a British Ministry of Technology Working Party report estimated that maintenance cost the United Kingdom (UK) was approximately £3000 million annually.[6,7]
- Annually, the cost of maintaining a military jet aircraft is around $1.6 million; approximately 11% of the total operating cost for an aircraft is spent on maintenance activities.[8]
- The typical size of a plant maintenance group in a manufacturing organization varied from 5 to 10% of the total operating force: in 1969, 1 to 17 persons, and in 1981, 1 to 12 persons.[9]
- The U.S. Department of Defense is the steward of the world's largest dedicated infrastructure, with a physical plant valued at approximately $570 billion on approximately 42,000 square miles of land, i.e., roughly the size of the state of Virginia.[10]
- The operation and maintenance budget request of the U.S. Department of Defense for fiscal year 1997 was on the order of $79 billion.[11]
- Annually, the U.S. Department of Defense spends around $12 billion for depot maintenance of weapon systems and equipment: Navy (59%), Air Force (27%), Army (13%), and others (1%).[10]
- In 1968, it was estimated that better maintenance practices in the U.K. could have saved approximately £300 million annually of lost production due to equipment unavailability.[12]

ENGINEERING MAINTENANCE IN THE 21ST CENTURY

Due to various factors, it was established in the previous century that "maintenance" must be an integral part of the production strategy for the overall success of an organization. For the effectiveness of the maintenance activity, the 21st century must build on this.[13]

It is expected that equipment of this century will be more computerized and reliable, in addition to being vastly more complex. Further computerization of equipment will significantly increase the importance of software maintenance, approaching, if not equal to, hardware maintenance. This century will also see more emphasis on maintenance with respect to such areas as the human factor, quality, safety, and cost effectiveness.

New thinking and new strategies will be required to realize potential benefits and turn them into profitability. All in all, profitable operations will be the ones that have employed modern thinking to evolve an equipment management strategy that takes effective advantage of new information, technology, and methods.

MAINTENANCE TERMS AND DEFINITIONS

This section presents some terms and definitions directly or indirectly used in engineering maintenance:[2,14–19]

- *Maintenance:* All actions appropriate for retaining an item/part/equipment in, or restoring it to, a given condition.
- *Maintenance engineering:* The activity of equipment/item maintenance that develops concepts, criteria, and technical requirements in conceptional and acquisition phases to be used and maintained in a current status during the operating phase to assure effective maintenance support of equipment.[14]
- *Preventive maintenance:* All actions carried out on a planned, periodic, and specific schedule to keep an item/equipment in stated working condition through the process of checking and reconditioning. These actions are precautionary steps undertaken to forestall or lower the probability of failures or an unacceptable level of degradation in later service, rather than correcting them after they occur.
- *Corrective maintenance:* The unscheduled maintenance or repair to return items/equipment to a defined state and carried out because maintenance persons or users perceived deficiencies or failures.
- *Predictive maintenance:* The use of modern measurement and signal-processing methods to accurately diagnose item/equipment condition during operation.
- *Maintenance concept:* A statement of the overall concept of the item/product specification or policy that controls the type of maintenance action to be employed for the item under consideration.
- *Maintenance plan:* A document that outlines the management and technical procedure to be employed to maintain an item; usually describes facilities, tools, schedules, and resources.
- *Reliability:* The probability that an item will perform its stated function satisfactorily for the desired period when used per the specified conditions.
- *Maintainability:* The probability that a failed item will be restored to adequately working condition.
- *Active repair time:* The component of downtime when repair persons are active to effect a repair.
- *Mean time to repair (MTTR):* A figure of merit depending on item maintainability equal to the mean item repair time. In the case of exponentially distributed times to repair, MTTR is the reciprocal of the repair rate.

- *Overhaul:* A comprehensive inspection and restoration of an item or a piece of equipment to an acceptable level at a durability time or usage limit.
- *Quality:* The degree to which an item, function, or process satisfies requirements of customer and user.
- *Maintenance person:* An individual who conducts preventive maintenance and responds to a user's service call to a repair facility, and performs corrective maintenance on an item. Also called custom engineer, service person, technician, field engineer, mechanic, repair person, etc.
- *Inspection:* The qualitative observation of an item's performance or condition.

MAINTENANCE PUBLICATIONS, ORGANIZATIONS, AND DATA INFORMATION SOURCES

This section presents selected publications, organizations, and data information sources directly or indirectly concerned with engineering maintenance.

PUBLICATIONS

Journals and Magazines

- *Journal of Quality in Maintenance Engineering,* MCB University Press, U.K.
- *Industrial Maintenance & Plant Operation,* Cahners Business Information, Inc., U.S.A.
- *Maintenance Technology,* Applied Technology Publications, Inc., U.S.A.
- *Maintenance Journal,* Engineer Information Transfer Pty. Ltd., Australia.
- *Reliability: The Magazine for Improved Plant Reliability,* Industrial Communications, Inc., U.S.A.
- *Maintenance and Asset Management Journal,* Conference Communications, Inc., U.K.

Books and Reports

- *Maintenance Engineering Handbook* edited by L.R. Higgins, McGraw-Hill Book Company, New York, 1988.
- *Engineering Maintenance Management* by B.W. Niebel, Marcel Dekker, Inc., New York, 1994.
- *Maintenance Fundamentals* by R.K. Mobley, Butterworth-Heinemann, Inc., Boston, 1999.
- *Maintenance Strategy* by A. Kelly, Butterworth-Heinemann, Inc., Oxford, U.K., 1997.
- *Reliability-Centered Maintenance* by J. Moubray, Industrial Press, Inc., New York, 1997.
- *Applied Reliability-Centered Maintenance* by J. August, Penn Well, Tulsa, Oklahoma, 1999.

- *Maintenance Planning and Control* by A. Kelly, Butterworth and Co. Ltd., London, 1984.
- *Quality, Warranty, and Preventive Maintenance* by I. Sahin and H. Polatoglu, Kluwer Academic Publishers, Boston, 1999.
- *Glossary of Reliability and Maintenance Terms* by T. McKenna and R. Oliverson, Gulf Publishing Company, Houston, Texas, 1997.
- *Maintenance Engineering Techniques, Report No. AMCP 706-132,* Department of the Army, Washington, D.C., 1975.
- *Guide to Reliability-Centered Maintenance, Report No. AMCP 705-2,* Department of the Army, Washington, D.C., 1985.
- *Queues, Inventories, and Maintenance* by P.M. Morse, John Wiley & Sons, New York, 1958.
- *Maintenance Engineering Handbook* by L.C. Morrow, McGraw-Hill Book Company, New York, 1966.
- *The Complete Handbook of Maintenance Management* by J.E. Heintzelman, Prentice-Hall, Inc., Englewood Cliffs, New Jersey, 1976.

ORGANIZATIONS

Professional

- Society for Maintenance and Reliability Professionals
  401 N. Michigan Ave., Chicago, Illinois, U.S.A.
- American Institute of Plant Engineers
  539 S. Lexington Pl., Anaheim, California, U.S.A.
- Society for Machinery Failure Prevention Technology
  4193 Sudley Road, Haymarket, Virginia, U.S.A.
- Maintenance Engineering Society of Australia (MESA)
  (A Technical Society of the Institution of Engineers, Australia)
  11 National Circuit, Barton, ACT, Australia
- Maintenance/Engineering Division, Canadian Institute of Mining, Metallurgy and Petroleum
  3400 de Maisonneuve Blvd West, Suite 1210, Montreal, Quebec, Canada
- The Institution of Plant Engineers
  77 Great Peter St., London, U.K.
- Japan Institute of Plant Maintenance
  Shuwa Shiba - Koen 3 - Chome Bldg.
  3-1-38, Shiba - Koen, Minato - Ku, Tokyo, Japan
- The Institute of Marine Engineers
  80 Coleman St., London, U.K.
- Society of Logistic Engineers
  8100 Professional Place, Suite 211, Hyattsville, Maryland, U.S.A.
- International Maintenance Institute
  P.O. Box 751896, Houston, Texas, U.S.A.
- Society of Automotive Engineers, Inc.
  400 Commonwealth Dr., Warrendale, Pennsylvania, U.S.A.

Consulting

- International Total Productive Maintenance (TPM) Institute, Inc.
  4018 Letort Lane, Allison Park, Pennsylvania, U.S.A.
- The Maintenance and Reliability Center
  506 East Stadium Hall, University of Tennessee, Knoxville,
  Tennessee, U.S.A.
- Maintenance and Housekeepers of Florida, Inc.
  750 S. Orange Blossom, Suite 106, Orlando, Florida, U.S.A.
- PM Maintenance Services
  RR5, Box 82-M, Georgetown, Delaware, U.S.A.
- Wolfson Maintenance
  Kilburn House Manchester Science Park, Pencroft Way,
  Manchester, U.K.
- Espinoza consulting
  P.O. Box 80935, Rochester, Michigan, U.S.A.
- Aladon Ltd.
  44 Regent Street, Lutterworth, Leicestershire, U.K.
- PM Safety Consultants Ltd.
  The Verdin Exchange, High Street, Winsford, Cheshire, U.K.
- Applied Reliability, Inc.
  11944 Justice Avenue, Suite E, Baton Rouge, Louisiana, U.S.A.
- BMT Reliability Consultants Ltd.
  12 Little Park Farm Road, Fareham, Hampshire, U.K.
- Bretech Engineering Ltd.
  70 Crown Street, P.O. Box 2331, Saint John, New Brunswick, Canada

DATA INFORMATION SOURCES

- GIDEP Data
  Government Industry Data Exchange Program (GIDEP) Operations Center
  Fleet Missile Systems, Analysis, and Evaluation Department of the Navy,
  Corona, California, U.S.A.
- National Technical Information Service (NTIS)
  5285 Port Royal Road, Springfield, Virginia, U.S.A.
- Defense Technical Information Center
  DTIC - FDAC
  8725 John J. Kingman Road, Suite 0944, Fort Belvoir, Virginia, U.S.A.
- Data on equipment used in electric power generation
  Equipment Reliability Information System (ERIS)
  Canadian Electrical Association,
  Montreal, Quebec, Canada
- Data on trucks and vans
  Commanding General
  Attn: DRSTA - QRA, U.S. Army
  Automotive - Tank Command,
  Warren, Michigan, U.S.A.

- Reliability Analysis Center
Rome Air Development Center,
Griffith Air Force Base, Rome, New York, U.S.A.

PROBLEMS

1. Discuss the needs for maintenance.
2. Define the following terms:
   - Maintenance
   - Maintenance engineering
3. What are the objectives of maintenance engineering?
4. What is the approximate amount of money spent annually on plant maintenance and operations by U.S. industry?
5. Write an essay on engineering maintenance in the 21st century.
6. What is the difference between preventive and predictive maintenance?
7. What is the difference between maintenance and maintainability?
8. List at least five sources for obtaining maintenance-related information.

REFERENCES

1. Zweekhorst, A., Evolution of maintenance, *Maintenance Technology,* October 1996, 9–14.
2. AMCP 706-132, *Engineering Design Handbook: Maintenance Engineering Techniques,* Department of Army, Washington, D.C., 1975.
3. Dhillon, B.S., Maintenance engineering, in *Reliability Engineering in Systems Design and Operation,* Van Nostrand Reinhold Co., New York, 1983, 239–273.
4. Dhillon, B.S., Maintenance, in *Reliability and Quality Control: Bibliography on General and Specialized Areas,* Beta Publishers, Gloucester, Ontario, Canada, 1992, 287–302.
5. Latino, C.J., *Hidden Treasure: Eliminating Chronic Failures Can Cut Maintenance Costs up to 60%,* Report, Reliability Center, Hopewell, Virginia, 1999.
6. Kelly, A., *Management of Industrial Maintenance,* Newnes-Butterworths, London, 1978.
7. Report by the Working Party on Maintenance Engineering, Department of Industry, London, 1970.
8. Kumar, U.D., New trends in aircraft reliability and maintenance measures, *Journal of Quality in Maintenance Engineering,* 5:4, 1999, 287–295.
9. Niebel, B.W., *Engineering Maintenance Management,* Marcel Dekker, New York, 1994.
10. Report on Infrastructure and Logistics, Department of Defense, Washington, D.C., 1995.
11. 1997 DOD Budget: Potential Reductions to Operation and Maintenance Program, United States General Accounting Office, Washington, D.C., 1996.
12. Kelly, A., *Maintenance Planning and Control,* Butterworths & Co., London, 1984.
13. Tesdahl, S.A. and Tomlingson, P.D., Equipment management breakthrough maintenance strategy for the 21st Century, in *Proceedings of the First International Conference on Information Technologies in the Minerals Industry,* December 1997, 39–58.

14. DOD INST.4151.12, *Policies Governing Maintenance Engineering within the Department of Defense,* Department of Defense, Washington D.C., June 1968.

15. Omdahl, T.P., editor, *Reliability, Availability and Maintainability (RAM) Dictionary,* ASQC Quality Press, Milwaukee, Wisconsin, 1988.

16. Mckenna, T. and Oliverson, R., *Glossary of Reliability and Maintenance Terms,* Gulf Publishing Co., Houston, Texas, 1997.

17. MIL-STD-721C, *Definitions of Terms for Reliability and Maintainability,* Department of Defense, Washington, D.C.

18. Von Alven, W.H., editor, *Reliability Engineering,* Prentice-Hall, Englewood Cliffs, New Jersey, 1964.

19. Naresky, J.J., Reliability definitions, *IEEE Transac. Reliability,* 19, 1970, 198–200.

2 Maintenance Mathematics

INTRODUCTION

As in the case of other engineering disciplines, mathematics is an indispensable maintenance tool. Mathematics[1] applications in engineering are relatively new. A history of mathematics is provided in Reference 1.

In maintenance, mathematics find applications in work sampling, inventory control analysis, failure data analysis, establishing optimum preventive maintenance policies, maintenance cost analysis, and project management control. Some of the areas of mathematics used in maintenance include set theory, probability, calculus, differential equations, Stochastic processes, and Laplace transforms. Even though many excellent texts are available in areas such as these, this chapter presents essential mathematical concepts to enable understanding of the material presented in the book. This should eliminate the need for readers to consult math books.

BOOLEAN ALGEBRA AND PROBABILITY PROPERTIES

Boolean algebra is important in probability theory and is named after George Boole (1813–1864), its originator.[2] Table 2.1 presents selective rules of Boolean algebra. The capital letters denote arbitrary sets or events and the symbol + denotes the union of sets or events. The intersection of sets in the table is written without the dot. Nevertheless, in some other documents it could have been written with the symbol $\cap$ or with a dot.

Important properties of probability are as follows:[3,4]

- The probability of occurrence of event, Y, is always

$$0 \leq P(Y) \leq 1 \tag{2.1}$$

where $P(Y)$ is the probability of occurrence of Y.

- The probability of occurrence and nonoccurrence of Y is given by

$$P(Y) + P(\overline{Y}) = 1 \tag{2.2}$$

where $\overline{Y}$ is the negation of Y and $P(\overline{Y})$ is the probability of nonoccurrence of Y.

TABLE 2.1
Commonly Used Boolean Algebra Rules[2,3]

Rule Description	Symbolism
Absorption law	$Y(Y + A) = Y$
	$Y + YA = Y$
Commutative law	$AY = YA$
	$A + Y = Y + A$
Idempotent law	$YY = Y$
	$Y + Y = Y$
Distributive law	$A(Y + B) = AY + AB$
	$A + (YB) = (A + Y)(A + B)$
Associative law	$A(YB) = (AY)B$
	$(A + B) + Y = A + (Y + B)$

- The probability of the sample space, S, is

$$P(S) = 1 \tag{2.3}$$

- The probability of the negation of the sample space S is

$$P(\bar{S}) = 1 \tag{2.4}$$

- The probability of an intersection of independent events, $Y_1, Y_2, Y_3 ..., Y_n$, is

$$P(Y_1 Y_2 Y_3 ... Y_n) = P(Y_1)P(Y_2)P(Y_3)...P(Y_n) \tag{2.5}$$

where
n = total number of events,
Y_i = ith event, for $i = 1, 2, 3,...,n$,
$P(Y_i)$ = probability of occurrence of event Y_i, for $i = 1, 2, 3,...,n$.

- The probability of the union of n independent events is given by

$$P(Y_1 + Y_2 + Y_3 + \cdots + Y_n) = 1 - \prod_{i=1}^{n}(1 - P(Y_i)) \tag{2.6}$$

- The probability of the union of n mutually exclusive events is expressed by

$$P(Y_1 + Y_2 + Y_3 + \cdots + Y_n) = \sum_{i=1}^{n} P(Y_i) \tag{2.7}$$

Note that for very small values of $P(Y_1), P(Y_2), P(Y_3),...,P(Y_n)$, Eq. (2.6) yields almost the same result to Eq. (2.7).

Example 2.1

Assume that in Eq. (2.6), we have $n = 2$, $P(Y_1) = .04$, and $P(Y_2) = .06$. Calculate the probability of the union of independent events Y_1 and Y_2. Use the same given data in Eq. (2.7) and comment on the results given by Eqs. (2.6) and (2.7).

For $n = 2$, Eq. (2.6) yields

$$P(Y_1 + Y_2) = 1 - \prod_{i=1}^{2}(1 - P(Y_i))$$

$$= P(Y_1) + P(Y_2) - P(Y_1)P(Y_2) \tag{2.8}$$

Substituting the given values for $P(Y_1)$ and $P(Y_2)$ into Eq. (2.8), we get

$$P(Y_1 + Y_2) = .04 + .06 - (.04)(.06)$$

$$= .0976$$

Using the same given data in Eq. (2.7) yields

$$P(Y_1 + Y_2) = P(Y_1) + P(Y_2)$$

$$= .1000$$

The above two results are almost identical.

PROBABILITY AND CUMULATIVE DISTRIBUTION FUNCTION DEFINITIONS

PROBABILITY

This is defined by[5]

$$P(Y) = \lim_{m \to \infty} (M/m) \tag{2.9}$$

where $P(Y)$ is the probability of occurrence of event Y and M is the total number of times that Y occurs in the m repeated experiments.

CUMULATIVE DISTRIBUTION FUNCTION

This is expressed by[5,6]

$$F(t) = \int_{-\infty}^{t} f(y)\, dy \tag{2.10}$$

where

    t = time,

    $F(t)$ = cumulative distribution function,

    $f(y)$ = probability density function.

By differentiating Eq. (2.10) with respect to t, we get

$$\frac{dF(t)}{dt} = \frac{d(\int_{-\infty}^{t} f(y)\,dy)}{dt} = f(t) \tag{2.11}$$

Setting $t = \infty$ in Eq. (2.10) yields

$$F(\infty) = \int_{-\infty}^{\infty} f(x)\,dx = 1 \tag{2.12}$$

This proves that the total area under the probability density curve is always equal to unity.

PROBABILITY DISTRIBUTIONS OF CONTINUOUS RANDOM VARIABLES

Over the years many continuous random variable probability distributions have been developed. This section presents some of those useful for performing mathematical maintenance analysis-related studies.[7-9]

EXPONENTIAL DISTRIBUTION

This is one of the most widely used probability distributions in engineering, particularly in reliability work.[10] It is relatively easy to handle in conducting analysis. The distribution probability density function is defined by

$$f(t) = \lambda e^{-\lambda t}, \quad t \ge 0, \ \lambda > 0 \tag{2.13}$$

where λ is the distribution parameter.

By substituting Eq. (2.13) into Eq. (2.10) we get the following expression for the exponential distribution cumulative distribution function:

$$F(t) = \int_{0}^{t} \lambda e^{-\lambda y}\,dy = 1 - e^{-\lambda t} \tag{2.14}$$

Example 2.2

By setting $t = \infty$ in Eq. (2.14) prove that the value of the cumulative distribution function is equal to unity.

Thus, for $t = \infty$ Eq. (2.14) becomes

$$F(\infty) = 1 - e^{-\lambda\infty} = 1 - 0 = 1$$

The above result proves that values of $F(t)$ for $t = \infty$ is always equal to unity.

RAYLEIGH DISTRIBUTION

This distribution, developed by John Rayleigh (1842–1919), is used often in reliability engineering and in the theory of sound.[1] Its probability density function is expressed by

$$f(t) = \left(\frac{2}{\alpha^2}\right) t e^{-(t/\alpha)^2}, \quad t \geq 0, \ \alpha > 0 \tag{2.15}$$

where α is the distribution parameter.

Inserting Eq. (2.15) into Eq. (2.10), we obtain

$$F(t) = 1 - e^{-(t/\alpha)^2} \tag{2.16}$$

The above equation is the Rayleigh distribution cumulative distribution function.

Example 2.3

Obtain an expression for the probability density function by using Eq. (2.16) in Eq. (2.11).

Substituting Eq. (2.16) into Eq. (2.11) yields

$$f(t) = \frac{dF(t)}{dt} = \left(\frac{2t}{\alpha^2}\right) e^{-(t/\alpha)^2} \tag{2.17}$$

Note that Eq. (2.17) is identical to Eq. (2.15). Thus, it proves that by differentiating the cumulative function, $F(t)$, with respect to time, t, yields the probability density function.

WEIBULL DISTRIBUTION

This distribution was developed by W. Weibull of the Royal Institute of Technology, Stockholm, in the early 1950s.[11] Weibull distribution is useful for representing many different physical phenomena. Its probability density function is defined by

$$f(t) = \frac{bt^{b-1} e^{-(t/\alpha)^b}}{\alpha^b}, \quad t \geq 0, \ b > 0, \ \alpha > 0 \tag{2.18}$$

where b and α are the shape and scale parameters, respectively.

Using Eq. (2.18) in Eq. (2.10), we get

$$F(t) = 1 - e^{-(t/\alpha)^b} \qquad (2.19)$$

Equation (2.19) is also known as Weibull cumulative distribution function.

Example 2.4

Obtain expressions by using Eq. (2.19) for $b = 1$ and $b = 2$ and comment on the resulting equations.

Thus, for $b = 1$ and $b = 2$ Eq. (2.19) yields the following expressions, respectively:

$$F(t) = 1 - e^{-t/\alpha} \qquad (2.20)$$

and

$$F(t) = 1 - e^{-(t/\alpha)^2} \qquad (2.21)$$

Equations (2.20) and (2.21) are identical to Eqs. (2.14) (i.e., for $1/\alpha = \lambda$) and (2.16), respectively. It means for $b = 1$ and $b = 2$ exponential and Rayleigh distributions are the special cases of the Weibull distribution, respectively.

NORMAL DISTRIBUTION

This distribution is sometime called the Gaussian distribution after Carl Friedrich Gauss (1777–1855), a German mathematician. It is one of the most widely used statistical distributions. The distribution probability density function is expressed by

$$f(t) = \frac{1}{\sigma\sqrt{2\pi}} \exp\left[-\frac{(t-\mu)^2}{2\sigma^2}\right], \quad -\infty < t < +\infty \qquad (2.22)$$

where μ and σ are the distribution parameters (i.e., mean and standard deviation, respectively).

Substituting Eq. (2.22) into Eq. (2.10), we obtain

$$F(t) = \frac{1}{\sigma\sqrt{2\pi}} \int_{-\infty}^{t} \exp\left[-\frac{(t-\mu)^2}{2\sigma^2}\right] dy \qquad (2.23)$$

The values of Eq. (2.23) are tabulated in various mathematical books.[4,6,11]

This distribution was actually discovered by De Moivre as early as in 1733 but due to historical error was attributed to Carl Gauss.[6]

GENERAL DISTRIBUTION

This distribution can represent a wide range of physical phenomena, and its probability density function is expressed by[12]

$$f(t) = [m\lambda s t^{s-1} + (1-m)\beta t^{\beta-1}\theta e^{\theta t^{\beta}}]\exp[-m\lambda t^{s} - (1-m)(e^{\theta t^{\beta}} - 1)] \quad (2.24)$$

$$\text{for } 0 \le m \le 1 \text{ and } \lambda, s, \beta, \theta > 0$$

where λ and θ are the scale parameters, β and s are the shape parameters.

By inserting Eq. (2.24) into Eq. (2.10), we get the following expression for the cumulative distribution function:

$$F(t) = 1 - \exp[-m\lambda t^{s} - (1-m)(e^{\theta t^{\beta}} - 1)] \quad (2.25)$$

The following statistical functions are the special cases of the general distribution:

- For $m = 1$: Weibull
- For $m = 1$ and $s = 2$: Rayleigh
- For $m = 1$ and $s = 1$: Exponential
- For $m = 0$ and $\beta = 1$: Extreme value
- For $s = 1$ and $\beta = 1$: Makeham
- For $s = 0.5$ and $\beta = 1$: Bathtub[13]

Table 2.2 presents cumulative distribution functions for the distributions discussed earlier.

TABLE 2.2
Cumulative Distribution Functions for Selective Distributions

Distribution Name	Cumulative Distribution Function ($F(t)$)
General	$1 - \exp[-m\lambda t^{s} - (1-m)(e^{\theta t^{\beta}} - 1)]$
Exponential	$1 - e^{-\lambda t}$
Rayleigh	$1 - e^{-(t/\alpha)^{2}}$
Weibull	$1 - e^{-(t/\alpha)^{b}}$
Normal	$\dfrac{1}{\sigma\sqrt{2\pi}} \displaystyle\int_{-\infty}^{t} \exp\left[-\dfrac{(t-\mu)^{2}}{2\sigma^{2}}\right] dy$

LAPLACE TRANSFORMS: INITIAL AND FINAL VALUE THEOREMS

Laplace transforms are useful for solving system of linear differential equations in mathematical maintenance analysis. These transforms are named for Pierre-Simon Laplace (1749–1827) who died exactly 100 years after the death of Isaac Newton.[1]

The Laplace transform of the function, $f(t)$, is expressed by[14-16]

$$f(s) = \int_0^\infty f(t)e^{-st} dt \tag{2.26}$$

where

    t = time,

    s = Laplace transform variable,

    $f(s)$ = Laplace transform of $f(t)$.

Example 2.5

Obtain Laplace transforms of the following two functions:

- $f(t) = 1$ (2.27)

- $f(t) = e^{-\lambda t}$ (2.28)

where λ is a constant.

Using Eq. (2.27) in Eq. (2.26), we get

$$f(s) = \int_0^\infty 1 \cdot e^{-st} dt = \frac{1}{s}, \quad \text{for } s > 0 \tag{2.29}$$

Substituting Eq. (2.28) into Eq. (2.26) yields

$$f(s) = \int_0^\infty e^{-\lambda t} e^{-st} dt = \left[\frac{-e^{-(s+\lambda)t}}{s+\lambda}\right]_0^\infty = \frac{1}{s+\lambda}, \quad \text{for } s > 0 \tag{2.30}$$

Laplace transforms of some selective functions are presented in Table 2.3.

INITIAL AND FINAL VALUE THEOREMS

The initial value theorem is given by

$$\lim_{t \to 0} f(t) \tag{2.31}$$

TABLE 2.3
Laplace Transforms of Some Common Functions

No.	$f(t)$	$f(s)$
1	$e^{-\lambda t}$	$\dfrac{1}{s+\lambda}$, for $s > -\lambda$
2	1	$\dfrac{1}{s}$, for $s > 0$
3	$\dfrac{df(t)}{dt}$	$s f(s) - f(0)$
4	$\int_0^t f(y)dy$	$\dfrac{f(s)}{s}$
5	t	$\dfrac{1}{s^2}$
6	$\dfrac{t^{m-1}}{(m-1)!}$	$\dfrac{1}{s^m}$, for $m = 1, 2, 3, \ldots$
7	$\dfrac{t^{m-1}e^{\lambda t}}{(m-1)!}$	$\dfrac{1}{(s-\lambda)^m}$, for $m = 1, 2, 3, \ldots$

The Laplace transform of Eq. (2.31) is[6]

$$\lim_{s \to \infty} sf(s) \tag{2.32}$$

The final value theorem is expressed by

$$\lim_{t \to \infty} f(t) \tag{2.33}$$

The Laplace transform of Eq. (2.33) is given by[6]

$$\lim_{s \to 0} sf(s) \tag{2.34}$$

Example 2.6

Prove that Eqs. (2.33) and (2.34) are equal. From Table 2.3 and Eq. (2.26) we write

$$L\left[\frac{df(t)}{dt}\right] = \int_0^\infty e^{-st}\frac{df(t)}{dt}\,dt = sf(s) - f(0) \tag{2.35}$$

where L is the Laplace transform operator.

The limit of

$$\int_0^\infty e^{-st}\frac{df(t)}{dt}\,dt$$

as $s \to 0$ is

$$\lim_{s \to 0} \int_0^\infty e^{-st} \frac{df(t)}{dt} \, dt = \int_0^\infty \frac{df(t)}{dt} \, dt$$

$$= \lim_{w \to \infty} \int_0^w \frac{df(t)}{dt} \, dt$$

$$= \lim_{w \to \infty} [f(w) - f(0)]$$

$$= \lim_{t \to \infty} f(t) - f(0) \tag{2.36}$$

The limit of $[s\, f(s) - f(0)]$ as $s \to 0$ is

$$\lim_{s \to 0} sf(s) - f(0) \tag{2.37}$$

From Eqs. (2.36) and (2.37) we obtain

$$\lim_{t \to \infty} f(t) - f(0) = \lim_{s \to 0} sf(s) - f(0) \tag{2.38}$$

Equation (2.38) yields

$$\lim_{t \to \infty} f(t) = \lim_{s \to 0} sf(s) \tag{2.39}$$

The above equation proves that Eqs. (2.33) and (2.34) are equal.

Example 2.7

Assume that we have

$$f(t) = \frac{\mu}{\lambda + \mu} + \frac{\lambda}{\lambda + \mu} e^{-(\lambda + \mu)t} \tag{2.40}$$

where λ and μ are parameters or constants and t is time.

Prove using Eq. (2.40) that Eqs. (2.33) and (2.34) yield identical results. By substituting Eq. (2.40) into Eq. (2.33), we get

$$\lim_{t \to \infty} \left[\frac{\mu}{\lambda + \mu} + \frac{\lambda}{\lambda + \mu} e^{-(\lambda + \mu)t} \right] = \frac{\mu}{\lambda + \mu} \tag{2.41}$$

Taking the Laplace transform of Eq. (2.40) yields

$$f(s) = \frac{\mu}{\lambda + \mu} \frac{1}{s} + \frac{\lambda}{\lambda + \mu} \frac{1}{(s + \lambda + \mu)}$$

$$= \frac{(s + \mu)}{s(s + \lambda + \mu)} \tag{2.42}$$

Inserting Eq. (2.42) into Eq. (2.34), we get

$$\lim_{s \to 0} s\left[\frac{(s+\mu)}{s(s+\lambda+\mu)}\right] = \frac{\mu}{\lambda+\mu} \tag{2.43}$$

Equations (2.41) and (2.43) prove that Eqs. (2.33) and (2.34) yield identical results.

ALGEBRAIC EQUATIONS

Mathematical maintenance analysis may involve determining roots of algebraic equations. A root may be described as a value of variable when insertion into the polynomial equation leads to the value of the equation equal to zero. When all roots of the polynomial equation are found, it is considered solved.[11,17,18]

QUADRATIC EQUATION

Although quadratic equations were solved around 2000 BC by Babylonians, in Western society before the seventeenth century the theory of equations was handicapped by the failure to recognize negative or complex numbers as the roots of equations.[1] A quadratic equation is defined by

$$Ax^2 + Bx + C = 0 \tag{2.44}$$

where x is a variable; A, B, and C are the constants.
    Solutions to Eq. (2.44) are given below:

$$x_1, x_2 = (-B \pm D^{1/2})/2A \tag{2.45}$$

where

$$D \equiv B^2 - 4AC \tag{2.46}$$

For real A, B, and C the roots can be classified as follows:

- For $D > 0$: real and unequal
- For $D = 0$: real and equal
- For $D < 0$: complex conjugate

If x_1 and x_2 are the roots of Eq. (2.44) then we have

$$x_1 x_2 = C/A \tag{2.47}$$

and

$$x_1 + x_2 = -B/A \tag{2.48}$$

CUBIC EQUATION

Italian mathematicians played an instrumental role in finding the algebraic solution to cubic equation. In 1545 Girolamo Cardano (1501–1576) published a Latin treatise on algebra at Nuremberg in Germany and included Tartaglia's solution of the cubic.[1] Cubic equation is expressed by

$$x^3 + B_1 x^2 + B_2 x + B_3 = 0 \tag{2.49}$$

where x is a variable; B_1, B_2, and B_3 are the constants.

Let

$$L = (3B_2 - B_1^2)/9 \tag{2.50}$$

$$M = (9B_1 B_2 - 27B_3 - 2B_1^3)/54 \tag{2.51}$$

$$N = [M + (L^3 + M^2)^{1/2}]^{1/3} \tag{2.52}$$

and

$$P = [M - (L^3 + M^2)^{1/2}]^{1/3} \tag{2.53}$$

The roots of Eq. (2.49) are given below:

$$x_1 = N + P - \frac{B_1}{3} \tag{2.54}$$

$$x_2 = -\frac{1}{2}(N + P) - \frac{B_1}{3} + \frac{1}{2}i\sqrt{3}(N - P) \tag{2.55}$$

$$x_3 = -\frac{1}{2}(N + P) - \frac{B_1}{3} - \frac{1}{2}i\sqrt{3}(N - P) \tag{2.56}$$

Let

$$T = L^3 + M^2 \tag{2.57}$$

For real B_1, B_2, and B_3 the roots can be classified as follows:

- For $T > 0$: one real and two complex conjugate
- For $T < 0$: all real and unequal
- For $T = 0$: all real and at least two equal

DIFFERENTIAL EQUATIONS

In mathematical maintenance analysis it may be necessary to find solutions to a set of linear differential equations, particularly when applying the Markov method. Even though there are various methods for solving differential equations, the Laplace transform approach is probably the most effective technique for solving a set of linear differential equations.

The following example demonstrates the application of Laplace transforms to solve a set of linear differential equations.

Example 2.8

Assume that the following two differential equations describe a repairable system:

$$\frac{dP_0(t)}{dt} = -\lambda P_0(t) + \mu P_1(t) \tag{2.58}$$

$$\frac{dP_1(t)}{dt} = -\mu P_1(t) + \lambda P_0(t) \tag{2.59}$$

where

$P_i(t)$ = probability that the system is in state i at time t, for $i = 0$ (working normally), $i = 1$ (failed),

λ = system failure rate,

μ = system repair rate.

At time $t = 0$, $P_0(0) = 1$, and $P_1(0) = 0$.

Prove by using Laplace transforms and Eqs. (2.58) and (2.59) that the probability of the system operating normally, i.e., $P_0(t)$, is given by Eq. (2.40).

Taking Laplace transforms of Eqs. (2.58) and (2.59), we get

$$sP_0(s) - P_0(0) = -\lambda P_0(s) + \mu P_1(s) \tag{2.60}$$

$$sP_1(s) - P_1(0) = -\mu P_1(s) + \lambda P_0(s) \tag{2.61}$$

where $P_i(s)$ is the Laplace transform of the probability that the system is in state i, for $i = 0, 1$.

For given initial conditions Eqs. (2.60) and (2.61) become

$$sP_0(s) - 1 = -\lambda P_0(s) + \mu P_1(s) \tag{2.62}$$

$$sP_1(s) = -\mu P_1(s) + \lambda P_0(s) \tag{2.63}$$

Rearranging Eq. (2.63) yields

$$P_1(s) = \frac{\lambda P_0(s)}{s + \mu} \tag{2.64}$$

Substituting Eq. (2.64) into Eq. (2.62), we obtain

$$P_0(s) = \frac{(s+\mu)}{s(s+\lambda+\mu)} \tag{2.65}$$

Taking the inverse Laplace transform of Eq. (2.65) results in

$$P_0(t) = \frac{\mu}{\lambda+\mu} + \frac{\lambda}{\lambda+\mu}e^{-(\lambda+\mu)t} \tag{2.66}$$

For $f(t) = P_0(t)$, Eqs. (2.40) and (2.66) are identical. It means Eq. (2.40) denotes the probability of the system operating normally when its (i.e., system) failure and repair rates are given.

PROBLEMS

1. Discuss the following Boolean algebra laws:
   - Idempotent law
   - Absorption law
2. Give a physical example of mutually exclusive events.
3. What are the independent events?
4. Define the following:
   - Probability density function
   - Cumulative distribution function
5. Prove that the cumulative distribution function of exponential distribution is given by

$$F(t) = 1 - e^{-\lambda t} \tag{2.67}$$

   where t is time and λ is the distribution parameter.
6. Write the probability density function of Weibull distribution. What are the special case distributions of the Weibull distribution?
7. Write the special case statistical functions of the general distribution.
8. Compare general and Weibull distributions.
9. Prove that the Laplace transform of $f(t) = t$ is given by

$$f(s) = \frac{1}{s^2} \tag{2.68}$$

10. Find the roots of the following equation:

$$x^3 + 2x^2 - 5x - 6 = 0 \tag{2.69}$$

   where x is a variable.

REFERENCES

1. Eves, H., *An Introduction to the History of Mathematics,* Holt, Rinehart, and Winston, New York, 1976.
2. Lipschutz, S., *Set Theory and Related Topics,* McGraw-Hill, New York, 1964.
3. NUREG-0492, *Fault Tree Handbook,* Nuclear Regulatory Commission (NRC), Washington, D.C., 1981.
4. Lipschutz, S., *Probability,* McGraw-Hill, New York, 1965.
5. Mann, N.R., Schafer, R.E., and Singpurwalla, N.D., *Methods for Statistical Analysis of Reliability and Life Data,* John Wiley & Sons, New York, 1974.
6. Shooman, M.L., *Probabilistic Reliability: An Engineering Approach,* McGraw-Hill, New York, 1968.
7. Patel, J.K., Kapadia, C.H., and Owen, D.B., *Handbook of Statistical Distributions,* Marcel Dekker, New York, 1976.
8. Ireson, W.G., editor, *Reliability Handbook,* McGraw-Hill, New York, 1966.
9. Dhillon, B.S., Life distributions, *IEEE Transac. Reliability,* 30, 1981, 457–460.
10. Davis, D.J., An analysis of some failure data, *J. Am. Stat. Assoc.,* 1952, 113–150.
11. Spiegel, M.R., *Mathematical Handbook of Formulas and Tables,* McGraw-Hill, New York, 1968.
12. Dhillon, B.S., A hazard rate model, *IEEE Transac. Reliability,* 29, 1979, 150.
13. Dhillon, B.S., *Reliability Engineering in Systems Design and Operation,* Van Nostrand Reinhold Co., New York, 1983.
14. Nixon, F.E., *Handbook of Laplace Transformation,* Prentice-Hall, Englewood Cliffs, New Jersey, 1965.
15. Spiegel, M.R., *Laplace Transforms,* McGraw-Hill, New York, 1965.
16. Oberhettinger, F. and Badii, L., *Tables of Laplace Transforms,* Springer-Verlag, Berlin, 1973.
17. Abramowitz, M. and Stegun, I.A., editors, *Handbook of Mathematical Functions,* U.S. Government Printing Office, Washington, D.C., 1972.
18. Burington, R.S., *Handbook of Mathematical Tables and Formulas,* McGraw-Hill, New York, 1973.

3 Maintenance Management and Control

INTRODUCTION

The management and control of maintenance activities are equally important to performing maintenance. Maintenance management may be described as the function of providing policy guidance for maintenance activities, in addition to exercising technical and management control of maintenance programs.[1,2] Generally, as the size of the maintenance activity and group increases, the need for better management and control become essential.

In the past, the typical size of a maintenance group in a manufacturing establishment varied from 5 to 10% of the operating force.[3] Today, the proportional size of the maintenance effort compared to the operating group has increased significantly, and this increase is expected to continue. The prime factor behind this trend is the tendency in industry to increase the mechanization and automation of many processes. Consequently, this means lesser need for operators but greater requirement for maintenance personnel.

There are many areas of maintenance management and control. This chapter presents some of the important ones.

MAINTENANCE DEPARTMENT FUNCTIONS AND ORGANIZATION

A maintenance department is expected to perform a wide range of functions including:[4-6]

- Planning and repairing equipment/facilities to acceptable standards
- Performing preventive maintenance; more specifically, developing and implementing a regularly scheduled work program for the purpose of maintaining satisfactory equipment/facility operation as well as preventing major problems
- Preparing realistic budgets that detail maintenance personnel and material needs
- Managing inventory to ensure that parts/materials necessary to conduct maintenance tasks are readily available
- Keeping records on equipment, services, etc.

- Developing effective approaches to monitor the activities of maintenance staff
- Developing effective techniques for keeping operations personnel, upper-level management, and other concerned groups aware of maintenance activities
- Training maintenance staff and other concerned individuals to improve their skills and perform effectively
- Reviewing plans for new facilities, installation of new equipment, etc.
- Implementing methods to improve workplace safety and developing safety education-related programs for maintenance staff
- Developing contract specifications and inspecting work performed by contractors to ensure compliance with contractual requirements

Many factors determine the place of maintenance in the plant organization including size, complexity, and product produced. The four guidelines useful in planning a maintenance organization are: establish reasonably clear division of authority with minimal overlap, optimize number of persons reporting to an individual, fit the organization to the personalities involved, and keep vertical lines of authority and responsibility as short as possible.[5]

One of the first considerations in planning a maintenance organization is to decide whether it is advantageous to have a centralized or decentralized maintenance function. Generally, centralized maintenance serves well in small- and medium-sized enterprises housed in one structure, or service buildings located in an immediate geographic area. Some of the benefits and drawbacks of centralized maintenance are as follows:[3]

Benefits

- More efficient compared to decentralized maintenance
- Fewer maintenance personnel required
- More effective line supervision
- Greater use of special equipment and specialized maintenance persons
- Permits procurement of more modern facilities
- Generally allows more effective on-the-job training

Drawbacks

- Requires more time getting to and from the work area or job
- No one individual becomes totally familiar with complex hardware or equipment
- More difficult supervision because of remoteness of maintenance site from the centralized headquarters
- Higher transportation cost due to remote maintenance work

In the case of decentralized maintenance, a maintenance group is assigned to a particular area or unit. Some important reasons for the decentralized maintenance are to reduce travel time to and from maintenance jobs, a spirit of cooperation between production and maintenance workers, usually closer supervision, and higher chances for maintenance personnel to become familiar with sophisticated equipment or facilities.

Past experience indicates that in large plants a combination of centralized and decentralized maintenance normally works best. The main reason is that the benefits of both the systems can be achieved with essentially a low number of drawbacks. Nonetheless, no one particular type of maintenance organization is useful for all types of enterprises.

MAINTENANCE MANAGEMENT BY OBJECTIVES, CRITICAL MAINTENANCE MANAGEMENT PRINCIPLES, AND MAINTENANCE PROGRAM EFFECTIVENESS EVALUATION QUESTIONS FOR MAINTENANCE MANAGERS

Improving a maintenance management program is a continuous process that requires progressive attitudes and active involvement. A nine-step approach for managing a maintenance program effectively is presented below:[6]

- *Identify existing deficiencies.* This can be accomplished through interviews with maintenance personnel and by examining in-house performance indicators.
- *Set maintenance goals.* These goals take into consideration existing deficiencies and identify targets for improvement.
- *Establish priorities.* List maintenance projects in order of savings or merit.
- *Establish performance measurement parameters.* Develop a quantifiable measurement for each set goal, for example, number of jobs completed per week and percentage of cost on repair.
- *Establish short- and long-range plans.* The short-range plan focuses on high-priority goals, usually within a one-year period. The long-range plan is more strategic in nature and identifies important goals to be reached within three to five years.
- *Document both long- and short-range plans and forward copies to all concerned individuals.*
- *Implement plan.*
- *Report status.* Preparing a brief report periodically, say semi-annually, and forward it to all involved individuals. The report contains for each objective identified in the short-range plan information on actual or potential slippage of the schedule and associated causes.
- *Examine progress annually.* Review progress at the end of each year with respect to stated goals. Develop a new short-range plan for the following year by considering the goals identified in the long-range plan and adjustments made to the previous year's planned schedule, resources, costs, and so on.

Over the years many maintenance management principles have been developed. Table 3.1 presents six critical maintenance management principles. These principles,

TABLE 3.1
Important Maintenance Management Principles

No.	Principle	Brief Description
1	Maximum productivity results when each involved person in an organization has a defined task to perform in a definitive way and a definite time.	This principle of scientific management formulated by Frederick W. Taylor in the late nineteenth century remains an important factor in management.[7,8]
2	Schedule control points effectively.	Schedule control points at intervals such that the problems are detected in time, thus the scheduled completion of the job is not delayed.
3	Measurement comes before control.	When an individual is given a definitive task to be accomplished using a good representative approach in a specified time, he/she becomes aware of management expectations. Control starts when managing supervisors compare the results against set goals.
4	The customer service relationship is the basis of an effective maintenance organization.	A good maintenance service is an important factor in maintaining facilities at an expected level effectively. The team approach fostered by the organizational setup is crucial to consistent, active control of maintenance activity.
5	Job control depends on definite, individual responsibility for each activity during the life span of a work order.	It is the responsibility of the maintenance department to develop, implement, and provide operating support for the planning and scheduling of maintenance work. It is the responsibility of the supervisory individuals to ensure proper and complete use of the system within their sphere of control.
6	The optimal crew size is the minimum number that can perform an assigned task effectively.	Most tasks require only one individual.

if applied on a regular basis, can help make a maintenance department productive and successful.[7]

The U.S. Energy Research and Development Administration conducted a study on maintenance management-related matters and formulated the following ten questions for maintenance managers to self-evaluate their maintenance effort:[8]

1. Are you aware of how your craftpersons spend their time; i.e., travel, delays, etc.?
2. Are you aware of what facility/equipment and activity consume most of the maintenance money?
3. Are you aware if the craftpersons use proper tools and methods to perform their tasks?

4. Have you balanced your spare parts inventory with respect to carrying cost vs. anticipated downtime losses?
5. With respect to job costs, are you in a position to compare the "should" with the "what"?
6. Do you ensure that maintainability factors are considered properly during the design of new or modified facilities/equipment?
7. Are you aware of how much time your foreman spends at the desk and at the job site?
8. Do you have an effective base to perform productivity measurements, and is productivity improving?
9. Are you aware of whether safety practices are being followed?
10. Are you providing the craftpersons with correct quality and quantity of material when and where they need it?

If an unqualified "yes" is the answer to each of the above questions, then your maintenance program is on a sound footing to meet organizational objectives. Otherwise, appropriate corrective measures are required.

ELEMENTS OF EFFECTIVE MAINTENANCE MANAGEMENT

There are many elements of effective maintenance management whose effectiveness is the key to the overall success of the maintenance activity. Many of these elements are described below.[6,8]

MAINTENANCE POLICY

A maintenance policy is one of the most important elements of effective maintenance management. It is essential for continuity of operations and a clear understanding of the maintenance management program, regardless of the size of a maintenance organization. Usually, maintenance organizations have manuals containing items such as policies, programs, objectives, responsibilities, and authorities for all levels of supervision, reporting requirements, useful methods and techniques, and performance measurement indices. Lacking such documentation, i.e., a policy manual, a policy document must be developed containing all essential policy information.

MATERIAL CONTROL

Past experience indicates that, on average, material costs account for approximately 30 to 40% of total direct maintenance costs.[8] Efficient utilization of personnel depends largely on effectiveness in material coordination. Material problems can lead to false starts, excess travel time, delays, unmet due dates, etc. Steps such as job planning, coordinating with purchasing, coordinating with stores, coordination of issuance of materials, and reviewing the completed job can help reduce material-related problems.

Deciding whether to keep spares in storage is one of the most important problems of material control. The subject of inventory control is discussed in detail in Chapter 7.

WORK ORDER SYSTEM

A work order authorizes and directs an individual or a group to perform a given task. A well-defined work order system should cover all the maintenance jobs requested and accomplished, whether repetitive or one-time jobs. The work order system is useful for management in controlling costs and evaluating job performance. Although the type and size of the work order can vary from one maintenance organization to another, a work order should at least contain information such as requested and planned completion dates, work description and its reasons, planned start date, labor and material costs, item or items to be affected, work category (preventive maintenance, repair, installation, etc.), and appropriate approval signatures.

EQUIPMENT RECORDS

Equipment records play a critical role in effectiveness and efficiency of the maintenance organization. Usually, equipment records are grouped under four classifications: maintenance work performed, maintenance cost, inventory, and files. The maintenance work performed category contains chronological documentation of all repairs and preventive maintenance (PM) performed during the item's service life to date. The maintenance cost category contains historical profiles and accumulations of labor and material costs by item. Usually, information on inventory is provided by the stores or accounting department. The inventory category contains information such as property number, size and type, procurement cost, date manufactured or acquired, manufacturer, and location of the equipment/item. The files category includes operating and service manuals, warranties, drawings, and so on.

Equipment records are useful when procuring new items/equipment to determine operating performance trends, troubleshooting breakdowns, making replacement or modification decisions, investigating incidents, identifying areas of concern, performing reliability and maintainability studies, and conducting life cycle cost and design studies.

PREVENTIVE AND CORRECTIVE MAINTENANCE

The basic purpose of performing PM is to keep facility/equipment in satisfactory condition through inspection and correction of early-stage deficiencies. Three principle factors shape the requirement and scope of the PM effort: process reliability, economics, and standards compliance.

A major proportion of a maintenance organization's effort is spent on corrective maintenance (CM). Thus, CM is an important factor in the effectiveness of maintenance organization. Both PM and CM are described in detail in Chapters 4 and 5.

JOB PLANNING AND SCHEDULING

Job planning is an essential element of the effective maintenance management. A number of tasks may have to be performed prior to commencement of a maintenance job; for example, procurement of parts, tools, and materials, coordination and delivery of parts, tools, and materials, identification of methods and sequencing, coordination with other departments, and securing safety permits.

Although the degree of planning required may vary with the craft involved and methods used, past experience indicates that on average one planner is required for every twenty craftpersons. Strictly speaking, formal planning should cover 100% of the maintenance workload but emergency jobs and small, straightforward work assignments are performed in a less formal environment. Thus, in most maintenance organizations 80 to 85% planning coverage is attainable.

Maintenance scheduling is as important as job planning. Schedule effectiveness is based on the reliability of the planning function. For large jobs, in particular those requiring multi-craft coordination, serious consideration must be given to using methods such as Program Evaluation and Review Technique (PERT) and Critical Path Method (CPM) to assure effective overall control. The CPM approach is described in detail later in this chapter.

BACKLOG CONTROL AND PRIORITY SYSTEM

The amount of backlog within a maintenance organization is one of the determining factors of maintenance management effectiveness. Identification of backlogs is important to balance manpower and workload requirements. Furthermore, decisions concerning overtime, hiring, subcontracting, shop assignments, etc., are largely based on backlog information. Management makes use of various indices to make backlog-related decisions.

The determination of job priority in a maintenance organization is necessary since it is not possible to start every job the day it is requested. In assigning job priorities, it is important to consider factors such as importance of the item or system, the type of maintenance, required due dates, and the length of time the job awaiting scheduling will take.

PERFORMANCE MEASUREMENT

Successful maintenance organizations regularly measure their performance through various means. Performance analyses contribute to maintenance department efficiency and are essential to revealing the downtime of equipment, peculiarities in operational behavior of the concerned organization, developing plans for future maintenance, and so on. Various types of performance indices for use by the maintenance management are discussed later in this chapter.

MAINTENANCE PROJECT CONTROL METHODS

Two widely used maintenance project control methods are Program Evaluation and Review Technique (PERT) and Critical Path Method (CPM). The development of PERT is associated with the U.S. Polaris project to monitor the effort of 250 prime contractors and 9000 subcontractors. PERT was the result of efforts of a team formed by the U.S. Navy's Special Project Office in 1958. Team members included the consulting firm of Booz, Allen, and Hamilton and the Lockheed Missile System Division.[9–12]

The history of CPM can be traced to 1956 when E.I. duPont de Nemours and Co. used a network model to schedule design and construction activities. The following year, CPM was used in the construction of a $10 million chemical plant in Louisville, Kentucky.

In maintenance and other projects three important factors of concern are time, cost, and resource availability. CPM and PERT deal with these factors individually and in combination.

PERT and CPM are similar. The major difference between the two is that when the completion times of activities of the project are uncertain, PERT is used and with the certainty of completion times, CPM is employed.[11]

The following steps are involved with PERT and CPM:[12]

- Break a project into individual jobs or tasks.
- Arrange these jobs/tasks into a logical network.
- Determine duration time of each job/task.
- Develop a schedule.
- Identify jobs/tasks that control the completion of project.
- Redistribute resources or funds to improve schedule.

The following sections present a formula to estimate activity expected duration times and CPM in detail.

ACTIVITY EXPECTED DURATION TIME ESTIMATION

The PERT scheme calls for three estimates of activity duration time using the following formula to calculate the final time:

$$T_a = \frac{OT + 4(MT) + PT}{6} \tag{3.1}$$

where
T_a = activity expected duration time,
OT = optimistic or minimum time an activity will require for completion,
PT = pessimistic or maximum time an activity will require for completion,
MT = most likely time an activity will require for completion. This is the time used for CPM activities.

Equation (3.1) is based on Beta distribution.[13]

Example 3.1

Assume that we have the following time estimates to accomplish an activity:

- OT = 55 days
- PT = 80 days
- MT = 60 days

Calculate the activity expected duration time.

Substituting the given data into Eq. (3.1), we get

$$T_a = \frac{55 + 4(60) + 80}{6} = 62.5 \text{ days}$$

The expected duration time for the activity is 62.5 days.

CRITICAL PATH METHOD (CPM)

Four symbols used to construct a CPM network are shown in Fig. 3.1. The circle denotes an event. Specifically, it represents an unambiguous point in the life of a project. An event could be the start or completion of an activity or activities, and usually the events are labeled by number. A circle shown with three divisions in Fig. 3.1(b) is also denotes an event. Its top half labels the event with a number, and the bottom portions indicate latest event time (LET) and earliest event time (EET). LET may be described as the latest time in which an event can be reached without delaying project completion. EET is the earliest time in which an activity can be accomplished or an event could be reached.

The continuous arrow represents an activity that consumes time, money, and manpower. This arrow always starts at a circle and ends at a circle. The dotted arrow denotes a dummy activity or a restraint. Specifically, this is an imaginary activity that does not consume time, money, or manpower. Figure 3.2 depicts an application of a dummy activity. It shows that activities L and M must be accomplished before activity N can start. However, only activity M must be completed prior to starting activity O.

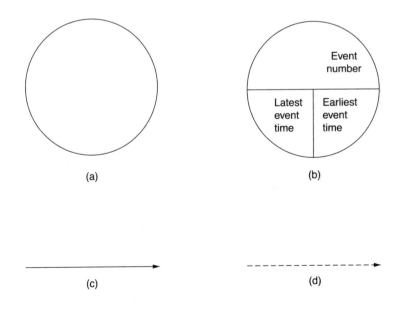

(a)

Event number

Latest event time | Earliest event time

(b)

(c)

(d)

FIGURE 3.1 CPM symbols: (a) circle, (b) circle with divisions, (c) continuous arrow, (d) dotted arrow.

TABLE 3.2

Maintenance Project Activities' Associated Data

Activity Identification	Immediate Predecessor Activity or Activities	Expected Duration in Days
L	–	12
M	–	2
N	L, M	2
O	L	6
P	O	3
S	N, P	9
T	S	15

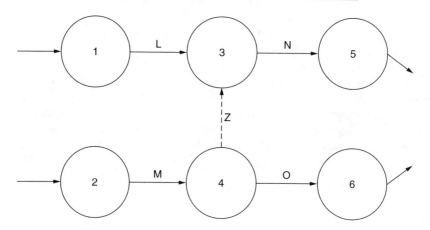

FIGURE 3.2 A portion of a CPM network with a dummy activity.

Example 3.2

A maintenance project was broken down into a set of seven activities, after which Table 3.2 was prepared. Prepare a CPM network using Fig. 3.1 symbols and Table 3.2 data, and determine the critical path associated with the network. A CPM network for given data in Table 3.2 is presented in Fig. 3.3.

In this figure, the following paths originate and terminate at events 1 and 7, respectively:

- M–N–S–T (2 + 2 + 9 + 15 = 28 days)
- L–X–N–S–T (12 + 0 + 2 + 9 + 15 = 38 days)
- L–O–P–S–T (12 + 6 + 3 + 9 + 15 = 45 days)

The quantities in parentheses above show the total time in days for each path. The dummy activity consumes zero time. By definition, the longest path through the network is the critical path. Inspection of the above three values shows that 45 days is

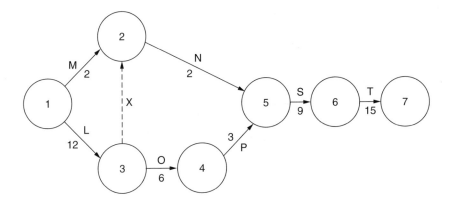

FIGURE 3.3 A CPM network for Table 3.2 data.

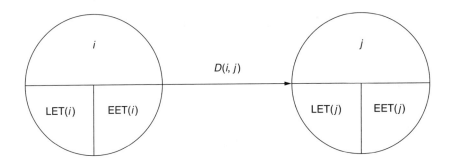

FIGURE 3.4 A single activity CPM network.

the largest time. Specifically, it will take 45 days from event 1 to reach event 7. Thus, this is the critical path. The word "critical" is used because any delay in the completion of activities along the critical path will result in delay of completion of the maintenance project.

Critical Path Determination Approach

For simple and straightforward CPM networks, the critical path can easily be identified in a manner discussed above. For complex networks a more systematic approach is required. This section presents one such approach with the aid of Fig. 3.4. The symbols used in the figure are defined below.

$EET(i)$ = earliest event time of event i
$EET(j)$ = earliest event time of event j
$LET(i)$ = latest event time of event i
$LET(j)$ = latest event time of event j
$D(i, j)$ = expected duration time of the activity between events i and j

The following steps are associated with the approach:

1. Construct CPM network.
2. Calculate EET of each event by making a forward pass of the network and using: For any event j,

$$EET(j) = \text{Maximum for all preceding}$$
$$i \text{ of } [EET(i) + D(i,j)] \tag{3.2}$$

Also,

$$EET(\text{first event}) = 0 \tag{3.3}$$

3. Calculate LET of each event by making a backward pass of the network and using: For any event i,

$$LET(i) = \text{Minimum for all succeeding}$$
$$j \text{ of } [LET(j) + D(i,j)] \tag{3.4}$$

Also,

$$LET(\text{last event}) = EET(\text{last event}) \tag{3.5}$$

If LET of all events of the network in question was calculated correctly, we should get

$$LET(\text{first event}) = 0 \tag{3.6}$$

4. Select network events with equal EET and LET. If the network results in only one path, i.e., from the first event to the last event, with EET = LET, this path is critical. Otherwise, go to next step.
5. Calculate the total float for each activity on each of the paths with EET = LET. The critical path is the path that results in the least sum of the total floats. The total float for any activity (i, j) can be calculated using the following equation:

$$\text{Total float} = LET(j) - EET(i) - D(i,j) \tag{3.7}$$

Example 3.3

Determine the critical path by calculating EET and LET of each event of the network shown in Fig. 3.3.

Using Eq. (3.2) we obtain EET of events 1, 2, 3, 4, 5, 6, and 7 as 0, 12, 12, 18, 21, 30, and 45, respectively. Similarly, with the aid of Eq. (3.4) the LET of events 1, 2, 3, 4, 5, 6, and 7 are 0, 19, 12, 18, 21, 30, and 45, respectively. Figure 3.5 shows

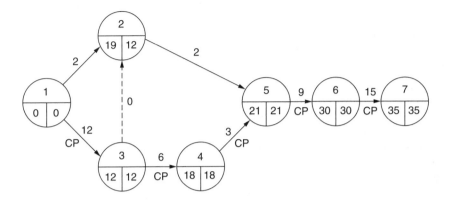

FIGURE 3.5 Redrawn Fig. 3.3 network with EETs and LETs.

a redrawn Fig. 3.3 CPM network with these EETs and LETs. The lower left quarter of each circle in Fig. 3.5 shows LET and the right right quarter the EET. The activities marked CP in Fig. 3.5 indicate the critical path as all the events that fall on this path have EET = LET, and it is the only path whose events have equal EETs and LETs. In other two paths the EET and LET of all events encountered are not equal.

CPM Advantages and Disadvantages

As with other methods, CPM has its advantages and disadvantages. Some of the CPM advantages are as follows:[14]

- It is an effective tool for monitoring project progress.
- It helps improve project understanding and communication among involved personnel.
- It highlights activities important to complete the project on time. These activities must be completed on time to accomplish the entire project on predicted time.
- It shows interrelationships in workflow and is useful in determining labor and resources needs in advance.
- It is an effective tool for controling costs and can easily be computerized.
- It helps avoiding duplications and omissions and determining project duration systematically.

Some of the disadvantages of the CPM are as follows:[14]

- Costly
- Time-consuming
- Poor estimates of activity times
- Inclination to use pessimistic estimates for activity times

MAINTENANCE MANAGEMENT CONTROL INDICES

Management employs various approaches to measure effectiveness of the maintenance function. Often it uses indices to manage and control maintenance. These indices show trends by using past data as a reference point. Usually, a maintenance organization employs various indices to measure maintenance effectiveness, as there is no single index that can accurately reflect the overall performance of the maintenance activity. The main objective of these indices is to encourage maintenance management to improve on past performance.

This section presents a number of broad and specific indices.[3,7,15,16] The broad indices indicate the overall performance of the organization with respect to maintenance and the specific indices indicate the performance in particular areas of the maintenance function. The values of all these indices are plotted periodically to show trends.

BROAD INDICATORS

This section presents three such indicators.

Index I

This is defined by

$$I_1 = \frac{TMC}{TS}$$ (3.8)

where
  TMC = total maintenance cost,
  TS   = total sales,
  I_1 = index parameter.

Past experience indicates that average expenditure for maintenance for all industry was around 5% of sales. However, there was a wide variation among industries. For example, the average values of I_1 for steel and chemical industries were 12.8 and 6.8%, respectively.

Index II

This is expressed by

$$I_2 = \frac{TMC}{TO}$$ (3.9)

where
  TO = total output expressed in gallons, tons, megawatts, etc.,
  I_2 = index parameter.

This index relates the total maintenance cost to the total output by the organization.

Index III

This is defined as follows:

$$I_3 = \frac{\text{TMC}}{\text{TIPE}} \tag{3.10}$$

where
I_3 = index parameter,
TIPE = total investment in plant and equipment.

This index relates the total maintenance cost to the total investment in plant and equipment. The approximate average figures for I_3 in the steel and chemical industries are 8.6 and 3.8%, respectively.

SPECIFIC INDICATORS

This section presents twelve such indicators.

Index IV

This is a useful index to control preventive maintenance activity within a maintenance organization and is defined by

$$I_4 = \frac{\text{TTPM}}{\text{TTEM}} \tag{3.11}$$

where
I_4 = index parameter,
TTPM = total time spent in performing preventive maintenance,
TTEP = total time spent for the entire maintenance function.

As per the past experience, the value of I_4 should be kept within 20 and 40% limits.

Index V

This index can be used to measure the accuracy of the maintenance budget plan and is expressed by

$$I_5 = \frac{\text{TAMC}}{\text{TBMC}} \tag{3.12}$$

where
I_5 = index parameter,
TAMC = total actual maintenance cost,
TBMC = total budgeted maintenance cost.

In this case, large variances indicate the need for immediate attention.

Index VI

This is a useful index for maintenance overhead control and is expressed by

$$I_6 = \frac{TMAC}{TMC} \tag{3.13}$$

where

I_6 = index parameter,
TMAC = total maintenance administration cost.

Index VII

This index is useful in scheduling work and is expressed as follows:

$$I_7 = \frac{PJCED}{TPJ} \tag{3.14}$$

where

I_7 = index parameter,
PJCED = total number of planned jobs completed by established due dates,
TPJ   = total number of planned jobs.

The value of I_7 should be high to keep backlogs down.

Index VIII

This index is useful in material control area and is defined by

$$I_8 = \frac{TPJAM}{TPJ} \tag{3.15}$$

where

I_8 = index parameter,
TPJAM = total number of planned jobs awaiting material.

Index IX

This index can be used to measure maintenance effectiveness and is defined by

$$I_9 = \frac{MHEUJ}{TMMH} \tag{3.16}$$

where

I_9 = index parameter,
MHEUJ = man-hours of emergency and unscheduled jobs,
TMMH = total maintenance man-hours worked.

Index X

This index can also be used to measure maintenance effectiveness and is expressed by

$$I_{10} = \frac{DTCB}{TDT} \tag{3.17}$$

where

I_{10} = index parameter,
DTCB = downtime caused by breakdowns,
TDT = total downtime.

Index XI

This is an important index used to measure inspection effectiveness and is defined by

$$I_{11} = \frac{NJI}{TIC} \tag{3.18}$$

where

I_{11} = index parameter,
NJI = number of jobs resulting from inspections,
TIC = total number of inspections completed.

Index XII

This index relates material and labor costs and is expressed by

$$I_{12} = \frac{TMLC}{TMMC} \tag{3.19}$$

where

I_{12} = index parameter,
TMLC = total maintenance labor cost,
TMMC = total maintenance materials cost.

Index XIII

This index relates maintenance cost to manufacturing cost and is defined by

$$I_{13} = \frac{TMC}{TMFC} \tag{3.20}$$

where

I_{13} = index parameter,
TMFC = total manufacturing cost.

Index XIV

This index relates maintenance cost to man-hours worked and is expressed by

$$I_{14} = \frac{\text{TMC}}{\text{TNMW}} \tag{3.21}$$

where

I_{14} = index parameter,
TNMW = total number of man-hours worked.

Index XV

This is a useful index to monitor progress in cost reduction efforts and is defined by

$$I_{15} = \frac{\text{PMMSJ}}{\text{MCPP}} \tag{3.22}$$

where

I_{15} = index parameter,
PMMSJ = percentage of maintenance man-hours spent on scheduled jobs,
MCPP = maintenance cost per unit of production.

PROBLEMS

1. List at least ten important functions of a maintenance department.
2. What are the advantages and disadvantages of centralized maintenance?
3. Describe a nine-step approach that can be used to manage a maintenance program.
4. Discuss six important maintenance management principles.
5. List ten questions the maintenance managers can use to self-evaluate effectiveness of their overall maintenance management.
6. Discuss the following three elements of effective maintenance management:
   - Maintenance policy
   - Work order system
   - Job planning
7. Describe the following three terms associated with CPM:
   - Dummy activity
   - Critical path
   - Total float
8. Determine the critical path of network shown in Fig. 3.5 by calculating total float for each activity.
9. What are the benefits and drawbacks of CPM?
10. Define two indices that can be used to evaluate overall performance of a maintenance organization.

REFERENCES

1. DOD Inst. 4151.12, *Policies Governing Maintenance Engineering within the Department of Defense,* Department of Defense, Washington, D.C., June 1968.
2. AMCP 706-132, *Engineering Design Handbook: Maintenance Engineering Techniques,* Department of the Army, Washington, D.C., June 1975.
3. Niebel, B.W., *Engineering Maintenance Management,* Marcel Dekker, New York, 1994.
4. Jordan, J.K., *Maintenance Management,* American Water Works Association, Denver, Colorado, 1990.
5. Higgins, L.R., *Maintenance Engineering Handbook,* McGraw-Hill, New York, 1988.
6. Dhillon, B.S., *Engineering Management,* Technomic Publishing Co., Lancaster, Pennsylvania, 1987.
7. Westerkamp, T.A., *Maintenance Manager's Standard Manual,* Prentice Hall, Paramus, New Jersey, 1997.
8. ERHQ-0004, *Maintenance Manager's Guide,* Energy Research and Development Administration, Washington, D.C., 1976.
9. Riggs, J.L. and Inoue, M.S., *Introduction to Operations Research and Management Science: A General Systems Approach,* McGraw-Hill, New York, 1974.
10. Lee, S.M., Moore, L.J., and Taylor, B.W., *Management Science,* Wm. C. Brown Co., Dubuque, Iowa, 1981.
11. Chase, R.B. and Aquilano, N.J., *Production and Operations Research: A Life Cycle Approach,* Richard D. Irwin, Homewood, Illinois, 1981.
12. Malcolm, D.G., Roseboom, J.H., Clark, C.E., and Fazar, W., Application of a technique for research and development program evaluation, *Operations Research,* 7, 1959, 646–669.
13. Clark, C.E., The PERT model for the distribution of an activity time, *Operations Research,* 10, 1962, 405–406.
14. Lomax, P.A., *Network Analysis: Applications to the Building Industry,* The English Universities Press Limited, London, 1969.
15. Hartmann, E., Knapp, D.J., Johnstone, J.J., and Ward, K.G., *How to Manage Maintenance,* American Management Association, New York, 1994.
16. Stoneham, D., *The Maintenance Management and Technology Handbook,* Elsevier Science, Oxford, 1998.

4 Preventive Maintenance

INTRODUCTION

Preventive maintenance (PM) is an important component of a maintenance activity. Within a maintenance organization it usually accounts for a major proportion of the total maintenance effort. PM may be described as the care and servicing by individuals involved with maintenance to keep equipment/facilities in satisfactory operational state by providing for systematic inspection, detection, and correction of incipient failures either prior to their occurrence or prior to their development into major failure.[1] Some of the main objectives of PM are to: enhance capital equipment productive life, reduce critical equipment breakdowns, allow better planning and scheduling of needed maintenance work, minimize production losses due to equipment failures, and promote health and safety of maintenance personnel.[2]

From time to time PM programs in maintenance organizations end up in failure (i.e., they lose upper management support) because their cost is either unjustifiable or they take a significant time to show results. It is emphasized that all PM must be cost-effective. The most important principle to keep continuous management support is: "If it is not going to save money, then don't do it!"

This chapter presents important aspects of PM.

PREVENTIVE MAINTENANCE ELEMENTS, PLANT CHARACTERISTICS IN NEED OF A PM PROGRAM, AND A PRINCIPLE FOR SELECTING ITEMS FOR PM

There are seven elements of PM as shown in Fig. 4.1.[1] Each element is discussed below.

1. *Inspection:* Periodically inspecting materials/items to determine their serviceability by comparing their physical, electrical, mechanical, etc., characteristics (as applicable) to expected standards
2. *Servicing:* Cleaning, lubricating, charging, preservation, etc., of items/ materials periodically to prevent the occurrence of incipient failures
3. *Calibration:* Periodically determining the value of characteristics of an item by comparison to a standard; it consists of the comparison of two instruments, one of which is certified standard with known accuracy, to detect and adjust any discrepancy in the accuracy of the material/parameter being compared to the established standard value

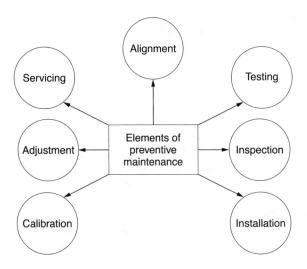

FIGURE 4.1 Elements of preventive maintenance.

4. *Testing:* Periodically testing or checking out to determine serviceability and detect electrical/mechanical-related degradation
5. *Alignment:* Making changes to an item's specified variable elements for the purpose of achieving optimum performance
6. *Adjustment:* Periodically adjusting specified variable elements of material for the purpose of achieving the optimum system performance
7. *Installation:* Periodic replacement of limited-life items or the items experiencing time cycle or wear degradation, to maintain the specified system tolerance

Some characteristics of a plant in need of a good preventive maintenance program are as follows:[2]

- Low equipment use due to failures
- Large volume of scrap and rejects due to unreliable equipment
- Rise in equipment repair costs due to negligence in areas such as regular lubrication, inspection, and replacement of worn items/components
- High idle operator times due to equipment failures
- Reduction in capital equipment expected productive life due to unsatisfactory maintenance

Table 4.1 presents 17 questions for determining the adequacy of a preventive maintenance program within an organization.

The answer "yes" or "no" to each question is given 5 or 0 points, respectively. A "maybe" answer is assigned a score from 1 to 4. A total score of less than 55 points indicates that the preventive maintenance program requires further improvements.[2]

TABLE 4.1
Preventive Maintenance Program Evaluation Questions

No.	Question	Yes (5 points)	Maybe (1–4 points)	No (0 points)
		Answer		
1	Is the trend in downtime recorded and reported regularly?			
2	Is there a formal PM program in place?			
3	Are inspectors performing their inspection duties full-time?			
4	Are check sheets controlled to assure 100% compliance?			
5	Are inspection routes developed/scheduled on the basis of work measurement methods?			
6	Are inspection reports randomly checked by supervisor to determine their accuracy?			
7	What percentage of downtime is due to maintenance?	($\geq 8\%$)	($8\% \leq$)	(Unknown)
8	Is the lubrication task performed through the scheduled usage of check sheets?			
9	Does maintenance management receive meaningful downtime reports?			
10	Is one individual responsible for the overall PM?			
11	Were lubrication routes developed and scheduled on the basis of time and method studies?			
12	Is data processing used to schedule and report PM inspections and lubrication?			
13	Are foreseeable problems, discovered through PM inspections, quickly reported?			
14	Is PM work highlighted in the cost-reporting system to permit routine analysis of PM as a distinct class of expenditure?			
15	Are lubrication requirements examined regularly to minimize the need for different types of lubricants?			
16	Is the analysis of breakdown reports performed to detect failure patterns that can be rectified by adjusting the PM program?			
17	Are plant/building assets examined regularly as an integral part of the formal inspection program?			

References 4 and 5 proposed the following principle or formula to be used when deciding to go ahead with a PM program:

$$(NB)(ACPBD)(\alpha) > CPMS \tag{4.1}$$

where
CPMS = total cost of preventive maintenance system,
α = a factor whose value is proposed to be taken as 70%; more specifically, 70% of the total cost of breakdowns,
NB   = number of breakdowns,
ACPBD = average cost per breakdown.

IMPORTANT STEPS FOR ESTABLISHING A PM PROGRAM

To develop an effective PM program, the availability of a number of items is necessary. Some of those items include accurate historical records of equipment, manufacturer's recommendations, skilled personnel, past data from similar equipment, service manuals, unique identification of all equipment, appropriate test instruments and tools, management support and user cooperation, failure information by problem/cause/action, consumables and replaceable components/parts, and clearly written instructions with a checklist to be signed off.[6]

There are a number of steps involved in developing a PM program. Figure 4.2 presents six steps for establishing a highly effective PM program in a short period. Each step is discussed below.[3]

1. *Identify and choose the areas.* Identify and selection of one or two important areas to concentrate the initial PM effort. These areas should be crucial to the success of overall plant operations and may be experiencing a high degree of maintenance actions. The main objective of this step is to obtain immediate results in highly visible areas, as well as to win concerned management support.

2. *Identify the PM needs.* Define the PM requirements. Then, establish a schedule of two types of tasks: daily PM inspections and periodic PM assignments. The daily PM inspections could be conducted by either maintenance or production personnel. An example of a daily PM inspection is to check the waste water settleable solids concentration. Periodic PM assignments usually are performed by the maintenance workers. Examples of such assignments are replacing throwaway filters, replacing drive belts, and cleaning steam traps and permanent filters.

3. *Establish assignment frequency.* Establish the frequency of the assignments. This involves reviewing the equipment condition and records. Normally, the basis for establishing the frequency is the experience of those familiar with the equipment and the recommendations of vendors and

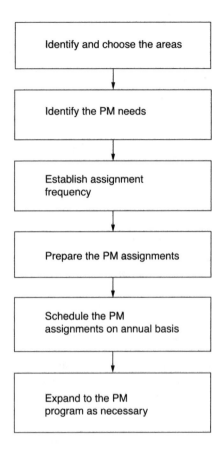

FIGURE 4.2 Six steps for developing a PM program.

engineering. It must be remembered that vendor recommendations are generally based on the typical usage of items under consideration.

4. *Prepare the PM assignments.* Daily and periodic assignments are identified and described in detail, then submitted for approval.

5. *Schedule the PM assignments on annual basis.* The defined PM assignments are scheduled on the basis of a twelve-month period.

6. *Expand the PM program as necessary.* After the implementation of all PM daily inspections and periodic assignments in the initially selected areas, the PM can be expanded to other areas. Experience gained from the pilot PM projects is instrumental to expanding the program.

PM MEASURES

Three important measures of PM are: mean preventive maintenance time (MPMT), median preventive maintenance time (MDPMT), and maximum preventive maintenance time (MXPMT). Each measure is described below.[1]

MEAN PREVENTIVE MAINTENANCE TIME (MPMT)

MPMT is the average item/equipment downtime needed to conduct scheduled PM. This time does not include PM time expended on the equipment/item during operation or administrative and logistic downtime.

Mean time for PM is defined by

$$\text{MPMT} = \frac{\sum_{i=1}^{m} f_i \text{MPMT}_i}{\sum_{i=1}^{m} f_i} \tag{4.2}$$

where

m = total number of data points,

MPMT_i = mean or average time needed to perform ith preventive maintenance action, for $i = 1, 2, 3,\ldots,m$,

f_i = frequency of ith preventive maintenance action in actions per operating hour after adjustment for equipment duty cycle.

MEDIAN PREVENTIVE MAINTENANCE TIME (MDPMT)

This is the item/equipment downtime needed to carry out 50% of all scheduled preventive maintenance actions on the item/equipment under the conditions outlined for MDPMT. For lognormal distributed PM times, the MDPMT is given by

$$\text{MDPMT} = \text{antilog}\left[\frac{\sum_{i=1}^{m} \lambda_i \log \text{MPMT}_i}{\sum_{i=1}^{m} \lambda_i}\right] \tag{4.3}$$

where

λ_i = constant failure rate of element i of the item/equipment for which maintainability is to be evaluated, adjusted for factors such as duty cycle, tolerance and interaction failures, and catastrophic failures that will lead to deterioration of item/equipment performance to the degree that a maintenance action will be started, for

$i = 1, 2, 3,\ldots,m.$

MAXIMUM PREVENTIVE MAINTENANCE TIME (MXPMT)

This is the maximum item/equipment downtime required to accomplish a given percentage of all scheduled preventive maintenance actions on the item/equipment under consideration. For lognormal distributed PM times, the MXPMT is given by

$$\text{MXPMT} = \text{antilog}(\log \text{MPMT}_m + y S_{\log \text{MPMT}}) \tag{4.4}$$

where
   y = value from table of normal distribution corresponding to the given percentage value at which MXPMT is defined (e.g., $y = 1.283$ for the 90th percentile and $y = 1.645$ for the 95th percentile). $\log \text{MPMT}_m$ is the mean of logarithms of MPMT_i and is expressed by

$$\log \text{MPMT}_m = \frac{\sum_{i=1}^{m} \lambda_i \log \text{MPMT}_i}{\sum_{i=1}^{m} \lambda_i} \tag{4.5}$$

$$S_{\log \text{MPMT}} = \left\{ \frac{\sum_{i=1}^{m} (\log \text{MPMT}_i)^2 - \left[\left(\sum_{i=1}^{m} \log \text{MPMT}_i \right)^2 \middle/ m \right]}{m - 1} \right\}^{1/2} \tag{4.6}$$

PM MODELS

Over the years many PM-related useful mathematical models have been developed. This chapter presents some of those models.

INSPECTION OPTIMIZATION MODEL I

Inspections are often disruptive, but they usually reduce downtime because of lesser number of failures. This model can be used to obtain the optimum number of inspections per facility per unit of time. Total facility downtime is defined by[7,8]

$$\text{TDT} = yT_i + \frac{cT_b}{y} \tag{4.7}$$

where
   TDT = total downtime per unit of time for a facility,
   c = a constant associated with a particular facility,
   T_b = facility downtime per breakdown or failure,
   T_i = facility downtime per inspection,
   y = number of inspections per facility per unit of time.

By differentiating Eq. (4.7) with respect to y, we get

$$\frac{d\,\text{TDT}}{dy} = T_i - \frac{cT_b}{y^2} \tag{4.8}$$

By setting Eq. (4.8) equal to zero and then rearranging, we obtain

$$y^* = \left(\frac{cT_b}{T_i}\right)^{1/2}$$

(4.9)

where
    y^* = optimum number of inspections per facility per unit of time.

By substituting Eq. (4.9) into Eq. (4.7) yields

$$\text{TDT}^* = 2(cT_iT_b)^{1/2}$$

(4.10)

where
    TDT = total optimal downtime per unit of time for a facility.

Example 4.1

An engineering facility was observed over a period of time and we obtained the following data:

$$T_b = 0.1 \text{ month}, \quad T_i = 0.05 \text{ month}, \quad c = 3$$

Using Eq. (4.9), calculate the optimal number of inspections per month.

    Using the given values in Eq. (4.9), we get

$$y^* = \left[\frac{3 \times 0.1}{0.05}\right]^{1/2} = 2.45 \text{ inspections per month}$$

The approximate number of optimal inspections per month is 2.

RELIABILITY AND MEAN TIME TO FAILURE DETERMINATION MODEL OF A SYSTEM WITH PERIODIC MAINTENANCE

This mathematical model can be used to calculate the reliability and mean time to failure of a system subject to periodic maintenance. The model is subject to the following assumptions:[9,10]

- A failed part is replaced with a new and statistically identical one.
- Periodic maintenance is performed on the system after every y hours, starting at time zero.

For periodic maintenance, the time interval of y hours is written as

$$y = iY + T, \quad i = 0, 1, 2,...; \quad 0 \leq T < Y$$

(4.11)

For $i = 1$ and $T = 0$, the reliability of a redundant system subject to periodic maintenance after every Y hours is given by

$$R_Y(y = Y) = R(Y) \tag{4.12}$$

For $i = 2$ and $T = 0$, we have

$$R_Y(y = 2Y) = [R(Y)]^2 \tag{4.13}$$

In this case, the system must operate the first Y hours without experiencing failure. Also, for another Y failure-free hours after the replacement of any failed part.

For $0 < T < Y$, another T hours of system failure-free operation is required. Thus,

$$R_Y(y = 2Y + T) = [R(Y)]^2 R(T) \tag{4.14}$$

In general form Eq. (4.14) is

$$R_Y(y = iY + T) = [R(Y)]^i [R(T)], \quad \text{for } i = 0, 1, 2, 3,...; \ 0 \le T < Y \tag{4.15}$$

The redundant system mean time to failure with the performance of periodic maintenance is given by

$$\text{MTTF}_{pm} = \int_0^\infty R_Y(y)dy \tag{4.16}$$

To evaluate Eq. (4.16), we write the integral over the range $0 < y < \infty$ as follows:

$$\text{MTTF}_{pm} = \sum_{i=0}^\infty \int_{iY}^{(i+1)Y} R_Y(y)dy \tag{4.17}$$

In Eq. (4.17), the integral of Eq. (4.16) is divided into time intervals of length Y. For $y = iY + T$, by substituting Eq. (4.15) into Eq. (4.17) we get

$$\text{MTTF}_{pm} = \sum_{i=0}^\infty \int_0^Y [R(Y)]^i R(T)dT \tag{4.18}$$

In Eq. (4.18) for $y = iY + T$, $dy = dT$ and the limits become 0 and Y.

Thus, rearranging Eq. (4.18) yields

$$\text{MTTF}_{pm} = \sum_{i=0}^\infty [R(Y)]^i \int_0^Y R(T)dT \tag{4.19}$$

Since

$$\sum_{i=0}^{\infty}[R(Y)]^i = \frac{1}{1-R(Y)} \tag{4.20}$$

Equation (4.19) becomes

$$\text{MTTF}_{pm} = \frac{\int_0^Y R(T)dT}{1-R(Y)} \tag{4.21}$$

Example 4.2

Assume that two independent and identical machines form a parallel system. Each machine's times to failure are exponentially distributed with a mean time to failure of 200 h. The periodic preventive maintenance (PM) is performed after every 100 h. Calculate the system mean time to failure with and without the performance of periodic PM.

Using the Chapter 12 information and the given data, the reliability of the two unit parallel system is

$$R(y) = 2e^{-y/200} - e^{-2y/200} \tag{4.22}$$

By substituting Eq. (4.22) into Eq. (4.21) yields

$$\begin{aligned}
\text{MTTF}_{pm} &= \frac{\int_0^{100} (2e^{-T/200} - e^{-2T/200})dt}{1-(2e^{-100/200} - e^{-2(100)/200})} \\
&= \frac{94.17}{0.1548} \\
&= 608.26 \text{ h}
\end{aligned}$$

By integrating Eq. (4.22) over the time interval $[0, \infty]$, we get system mean time to failure without the performance of periodic maintenance as follows:

$$\text{MTTF}_{pm} = \int_0^{\infty} (2e^{-y/200} - e^{-2y/200})dy = 300 \text{ h}$$

This means that periodic PM helped improve the system mean time to failure from 300 to 608.26 h.

INSPECTION OPTIMIZATION MODEL II

This is similar to Inspection Frequency Model I. It can be used to determine optimum inspection frequency in order to minimize the per-unit-of-time equipment/facility downtime. In this model facility/equipment (per-unit time) total downtime is the function of inspection frequency. Mathematically, it is defined as follows:[11,12]

$$TDT(n) = DT_r + DT_i$$

$$= \frac{\lambda(n)}{\mu} + \frac{n}{\theta} \qquad (4.23)$$

where

$TDT(n)$ = facility/equipment total downtime per unit of time,
DT_i = equipment/facility downtime due to per-unit-of-time inspection,
DT_r = equipment/facility downtime due to per-unit-of-time repairs,
n = inspection frequency,
$\lambda(n)$ = equipment/facility failure rate,
μ = equipment/facility repair rate,
$1/\theta$ = mean of exponentially distributed inspection times.

By differentiating Eq. (4.23) with respect to n, we get

$$\frac{d\,TDT(n)}{dn} = \frac{d\,\lambda(n)}{dn}\frac{1}{\mu} + \frac{1}{\theta} \qquad (4.24)$$

Setting Eq. (4.24) equal to zero and rearranging yields

$$\frac{d\lambda(n)}{dn} = -\frac{\mu}{\theta} \qquad (4.25)$$

The value of n will be optimum when the left and right sides of Eq. (4.25) are equal. At this point the equipment/facility total downtime will be minimal.

Example 4.3

Assume the failure rate of a system is defined by

$$\lambda(n) = fe^{-n} \qquad (4.26)$$

where f is the system failure rate at $n = 0$. Obtain an expression for the optimal value of n by using Eq. (4.25).

By substituting Eq. (4.26) into Eq. (4.25), we get

$$-fe^{-n} = -\frac{\mu}{\theta} \qquad (4.27)$$

Rearranging Eq. (4.27) yields

$$n^* = \ln\left[\frac{f\theta}{\mu}\right] \qquad (4.28)$$

where

  n^* = optimal inspection frequency.

Example 4.4

Assume that in Example 4.3 we have the following:

$$\frac{1}{\mu} = 0.02 \text{ month}, \quad \frac{1}{\theta} = 0.005 \text{ month}, \quad f = 1 \text{ failure per month}$$

Calculate the optimal value of the inspection frequency, n.

  Inserting the given values into Eq. (4.28) yields

$$n^* = \ln\left[\frac{1 \times 0.02}{0.005}\right] = 1.39 \text{ inspections per month}$$

This means that roughly one inspection per month will be optimal.

INSPECTION OPTIMIZATION MODEL III

This is a useful mathematical model that can be used to calculate optimum inspection frequency to maximize profit. The model is developed on the premise that the facility/equipment under repair lead to zero output, thus less profit. Furthermore, if equipment is inspected too often, there is danger that it may be more costly due to factors such as loss of production, cost of materials, and wages than losses due to breakdowns. The following assumptions are associated with this model:[11,12]

- The equipment failure rate is a function of inspections.
- Times to inspection are exponentially distributed.
- Equipment failure and repair rates are constant.

  The following symbols were used to develop equations for the model:

n = number of inspections performed per unit of time,
$1/\theta$ = mean of exponentially distributed inspection times,
p = profit at no downtime losses,
C_i = average inspection cost per uninterrupted unit of time,
C_r = average cost of repair per uninterrupted unit of time,
λ = equipment failure rate,
μ = equipment repair rate.

  Profit per unit of time is expressed by:[11,12]

$$PR = p - PL_i - PL_r - IC - RC$$

$$= p - \frac{pn}{\theta} - \frac{p\lambda(n)}{\mu} - \frac{nc_i}{\theta} - \frac{C_r\lambda(n)}{\mu} \tag{4.29}$$

where
   PL_i = production output value loss per unit of time due to inspections,
   PL_r = production output value loss per unit of time due to repairs,
   IC = inspection cost per unit of time,
   RC = repair cost per unit of time.

By differentiating Eq. (4.29) with respect to n and then equating it to zero yield

$$\frac{d\,\text{PR}}{dn} = -\frac{p}{\theta} - \frac{p}{\mu}\frac{d\lambda(n)}{dn} - \frac{C_i}{\theta} - \frac{C_r}{\mu}\frac{d\lambda(n)}{dn} = 0 \tag{4.30}$$

Rearranging Eq. (3.30), we get

$$\frac{d\lambda(n)}{dn} = -\left[\frac{1}{\theta}(p + C_i)\right] \bigg/ \left(\frac{p}{\mu} + \frac{C_r}{\mu}\right) \tag{4.31}$$

The value of n will be optimal when left and right sides of Eq. (4.31) are equal. At this point, the profit will be at its maximum value.

Example 4.5

Assume the failure rate of a manufacturing system is defined by Eq. (4.26) in Example 4.3. Develop an expression for the optimal value of n with the aid of Eq. (4.31).

Using Eq. (4.26) in Eq. (4.31) yields

$$-fe^{-n} = -\left[\frac{1}{\theta}(p + C_i)\right] \bigg/ \left(\frac{p}{\mu} + \frac{C_r}{\mu}\right) \tag{4.32}$$

By rearranging Eq. (4.32), we get

$$n^* = \ln\left[\frac{f\theta(p + C_r)}{\mu(p + C_i)}\right] \tag{4.33}$$

where
   n^* = optimal manufacturing system inspection frequency.

Example 4.6

Suppose that in Example 4.5 we have the following data:

   p = \$10,000 per month
   f = 2 failures per month
   $1/\mu$ = 0.04 month
   $1/\theta$ = 0.01 month
   C_i = \$75 per month
   C_r = \$400 per month

Determine the optimal value of n by using Eq. (4.33).

By inserting the specified data values into Eq. (4.33), we obtain

$$n^* = \ln\left[\frac{2 \times 0.04 \times (10,000 + 400)}{0.01 \times (10,000 + 75)}\right]$$

$$= 2.11 \text{ inspections per month}$$

For optimal performance, approximately two inspections per month should be performed.

PM MARKOV MODEL

This mathematical model represents a system that can either fail completely or undergo periodic PM. The failed system is repaired. The system transition diagram is shown in Fig. 4.3. The model is useful to predict system availability, probability of system down for PM, and probability of system failure.

The following assumptions are associated with the model:

- System PM, failure, and repair rates are constant.
- After repair or PM the system is as good as new.

The following symbols were used to develop equations for the model:

j = the jth system state, $j = 0$ (system operating normally), $j = 1$ (system failed), $j = p$ (system down for PM),
$P_j(t)$ = probability that the system is in state j at time t, for $j = 0, 1, p$,
λ = system failure rate,
μ = system repair rate,
λ_p = rate of system down for PM,
μ_p = rate of system PM performance.

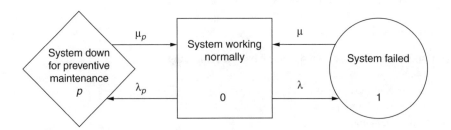

FIGURE 4.3 System transition diagram.

Using the Markov method we write the following equations for Fig. 4.3:[13]

$$\frac{dP_0(t)}{dt} + (\lambda + \lambda_p) = \mu P_1(t) + \mu_p P_p(t) \tag{4.34}$$

$$\frac{dP_p(t)}{dt} + \mu_p P_p(t) = \lambda_p P_0(t) \tag{4.35}$$

$$\frac{dP_1(t)}{dt} + \mu P_1(t) = \lambda P_0(t) \tag{4.36}$$

At time $t = 0$, $P_0(0) = 1$ and $P_p(0) = P_1(0) = 0$.

Solving Eqs. (4.34)–(4.36), we get

$$P_0(t) = \frac{\mu_p \mu}{m_1 m_2} + \left[\frac{(m_1 + \mu_p)(m_1 + \mu)}{m_1(m_1 - m_2)}\right]e^{m_1 t} - \left[\frac{(m_2 + \mu_p)(m_2 + \mu)}{m_2(m_1 - m_2)}\right]e^{m_2 t} \tag{4.37}$$

$$P_1(t) = \frac{\lambda \mu_p}{m_1 m_2} + \left[\frac{\lambda m_1 + \lambda \mu_p}{m_1(m_1 - m_2)}\right]e^{m_1 t} - \left[\frac{(\mu_p + m_2)\lambda}{m_2(m_1 - m_2)}\right]e^{m_2 t} \tag{4.38}$$

$$P_p(t) = \frac{\lambda_p \mu}{m_1 m_2} + \left[\frac{\lambda_p m_1 + \lambda_p \mu}{m_1(m_1 - m_2)}\right]e^{m_1 t} - \left[\frac{(\mu + m_2)\lambda_p}{m_2(m_1 - m_2)}\right]e^{m_2 t} \tag{4.39}$$

where

$$m_1, m_2 = \frac{-B \pm [B^2 - 4(\mu_p \mu + \lambda \mu_p + \lambda_p \mu)]^{1/2}}{2} \tag{4.40}$$

$$B \equiv (\mu + \mu_p + \lambda + \lambda_p) \tag{4.41}$$

$$m_1 + m_2 = -B \tag{4.42}$$

$$m_1 m_2 = \mu_p \mu + \lambda_p \mu + \lambda \mu_p \tag{4.43}$$

The probability of system failure is given by Eq. (4.38), the probability of system down for PM by Eq. (4.39), and the system availability by Eq. (4.37). As time t becomes large in Eq. (4.37), we get the following expression for the system steady-state availability:

$$AV_{ss} = \frac{\mu \mu_p}{\mu_p \mu + \lambda_p \mu + \lambda \mu_p} \tag{4.44}$$

Example 4.7

Assume that in Eq. (4.44) we have $\lambda = 0.005$ failures per hour, $\lambda_p = 0.008$ per hour, $\mu = 0.009$ repairs per hour, and $\mu_p = 0.009$ per hour. Calculate the system steady-state availability.

Substituting the given values into Eq. (4.44) yields

$$AV_{SS} = \frac{0.009 \times 0.009}{(0.009 \times 0.009) + (0.008 \times 0.009) + (0.005 \times 0.009)}$$

$$= 0.4091$$

There is an approximately 41% chance that the system will be available for service. Specifically, the system steady-state availability is 41%.

PM ADVANTAGES AND DISADVANTAGES

The performance of PM has many advantages including increase in equipment availability, performed as convenient, balanced workload, reduction in overtime, increase in production revenue, consistency in quality, reduction in need for standby equipment, stimulation in preaction instead of reaction, reduction in parts inventory, improved safety, standardized procedures, times, and costs, scheduled resources on hand, and useful in promoting benefit/cost optimization.[4,6]

Some disadvantages of PM are: exposing equipment to possible damage, using a greater number of parts, increases in initial costs, failures in new parts/components, and demands more frequent access to equipment/item.[6]

PROBLEMS

1. Discuss at least five important elements of PM.
2. What are the symptoms of a plant in need of a good PM program?
3. What are the important questions that can be asked to determine the adequacy of a PM program?
4. Comment on the principle or formula proposed to decide whether to go ahead with a PM program.
5. List at least ten items whose availability is essential to develop an effective PM program.
6. Discuss important steps for developing a PM program.
7. Discuss the following:
   - Mean PM time
   - Maximum PM time
   - Median PM time
8. Three independent and identical machines form a parallel system. Each machine's times to failure are exponentially distributed with a mean time to failure of 150 h. The periodic PM is performed after every 75 h.

Determine the system mean time to failure with and without performance of periodic PM.

9. Obtain steady-state expressions for Eqs. (4.38) and (4.39). Obtain an expression for the system steady-state unavailability.

10. What are the benefits and drawbacks of performing PM?

REFERENCES

1. AMCP 706-132, *Engineering Design Handbook: Maintenance Engineering Techniques,* Department of Defense, Washington, D.C., 1975.

2. Niebel, B.W., *Engineering Maintenance Management,* Marcel Dekker, New York, 1994.

3. Westerkamp, T.A., *Maintenance Manager's Standard Manual,* Prentice-Hall, Paramus, New Jersey, 1997.

4. Levitt, J., Managing preventing maintenance, *Maintenance Technology,* February 1997, 20–30.

5. Levitt, J., *Maintenance Management,* Industrial Press, New York, 1997.

6. Patton, J.D., *Preventive Maintenance,* Instrument Society of America, Research Triangle Park, North Carolina, 1983.

7. Wild, R., *Essentials of Production and Operations Management,* Holt, Rinehart, and Winston, London, 1985.

8. Dhillon, B.S., *Mechanical Reliability: Theory, Models, and Applications,* American Institute of Aeronautics and Astronautics, Washington, D.C., 1988.

9. Von Alven, W.H., *Reliability Engineering,* Prentice-Hall, Englewood Cliffs, New Jersey, 1964.

10. Rosenheim, D.E., Analysis of reliability improvement through redundancy, in *Proceedings of the New York University Conference on Reliability Theory,* New York University, New York, June 1958, 119–142.

11. Jardine, A.K.S., *Maintenance, Replacement and Reliability,* Pitman Publishing, London, 1973.

12. Dhillon, B.S., *Systems Reliability, Maintainability, and Management,* Petrocelli Books, New York, 1983.

13. Dhillon, B.S., *Reliability Engineering in Systems Design and Operation,* Van Nostrand Reinhold Co., New York, 1983.

5 Corrective Maintenance

INTRODUCTION

Although every effort is made to make engineering systems as reliable as possible through design, preventive maintenance, and so on, from time to time they do fail. Consequently, they are repaired to their operational state. Thus, repair or corrective maintenance is an important component of maintenance activity. Corrective maintenance may be defined as the remedial action carried out due to failure or deficiencies discovered during preventive maintenance, to repair an equipment/item to its operational state.[1-3]

Usually, corrective maintenance is an unscheduled maintenance action, basically composed of unpredictable maintenance needs that cannot be preplanned or programmed on the basis of occurrence at a particular time. The action requires urgent attention that must be added, integrated with, or substituted for previously scheduled work items. This incorporates compliance with "prompt action" field changes, rectification of deficiencies found during equipment/item operation, and performance of repair actions due to incidents or accidents.

A substantial part of overall maintenance effort is devoted to corrective maintenance, and over the years many individuals have contributed to the area of corrective maintenance. This chapter presents some important aspects of corrective maintenance.

CORRECTIVE MAINTENANCE TYPES

Corrective maintenance may be classified into five major categories as shown in Fig. 5.1.[1,4] These are: fail-repair, salvage, rebuild, overhaul, and servicing. These categories are described below.

1. *Fail-repair:* The failed item is restored to its operational state.
2. *Salvage:* This element of corrective maintenance is concerned with disposal of nonrepairable material and use of salvaged material from nonrepairable equipment/item in the repair, overhaul, or rebuild programs.
3. *Rebuild:* This is concerned with restoring an item to a standard as close as possible to original state in performance, life expectancy, and appearance. This is achieved through complete disassembly, examination of all components, repair and replacement of worn/unserviceable parts as per original specifications and manufacturing tolerances, and reassembly and testing to original production guidelines.

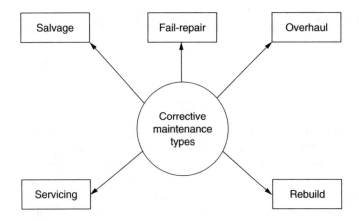

FIGURE 5.1 Types of corrective maintenance.

4. *Overhaul:* Restoring an item to its total serviceable state as per maintenance serviceability standards, using the "inspect and repair only as appropriate" approach.

5. *Servicing:* Servicing may be needed because of the corrective maintenance action, for example, engine repair can lead to crankcase refill, welding on, etc. Another example could be that the replacement of an air bottle may require system recharging.

CORRECTIVE MAINTENANCE STEPS, DOWNTIME COMPONENTS, AND TIME REDUCTION STRATEGIES AT SYSTEM LEVEL

Different authors have laid down different sequential steps for performing corrective maintenance. For example, Reference 2 presents nine steps (as applicable): localize, isolate, adjust, disassemble, repair, interchange, reassemble, align, and checkout. Reference 3 presents seven steps (as applicable): localization, isolation, disassembly, interchange, reassemble, alignment, and checkout.

For our purpose, we assume that corrective maintenance is composed of five major sequential steps, as shown in Fig. 5.2.[1] These steps are: fault recognition, localization, diagnosis, repair, and checkout.

The major corrective maintenance downtime components are active repair time, administrative and logistic time, and delay time.[1,5] The active repair time is made up of the following subcomponents:

- Preparation time
- Fault location time
- Spare item obtainment time
- Fault correction time
- Adjustment and calibration time
- Checkout time

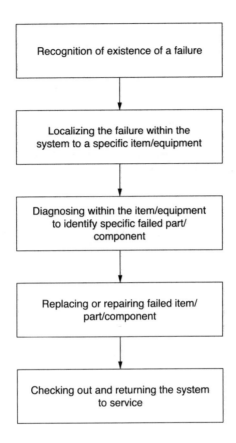

FIGURE 5.2 Corrective maintenance sequential steps.

Reduction in corrective maintenance time is useful to improve maintenance effectiveness. Some strategies for reducing the system-level corrective maintenance time are as follows:[6]

- *Efficiency in fault recognition, location, and isolation:* Past experience indicates that in electronic equipment, fault isolation and location consume the most time within a corrective maintenance activity. In the case of mechanical items, often the largest contributor is repair time. Factors such as well-designed fault indicators, good maintenance procedures, well-trained maintenance personnel, and an unambiguous fault isolation capability are helpful in lowering corrective maintenance time.
- *Effective interchangeability:* Good physical and functional interchangeability is useful in removing and replacing parts/items, reducing maintenance downtime, and creating a positive impact on spares and inventory needs.
- *Redundancy:* This is concerned with designing in redundant parts that can be switched in at the moment of need so the equipment/system continues to operate while the faulty part is being repaired. In this case the overall

maintenance workload may not be reduced, but the equipment/system downtime could be impacted significantly.

- *Effective accessibility:* Often a significant amount of time is spent accessing the failed part. Proper attention to accessibility during design can help reduce part accessibility time and, in turn, the corrective maintenance time.
- *Human factor considerations:* Attention paid to human factors during design in areas such as readability of instructions, size, shape, and weight of components, selection and placement of dials and indicators, size and placement of access, gates, and readability, and information processing aids can help reduce corrective maintenance time significantly.

CORRECTIVE MAINTENANCE MEASURES

There are various measures associated with corrective maintenance. This section presents three such measures.[1,6-8]

MEAN CORRECTIVE MAINTENANCE TIME

This is defined by

$$T_{mcm} = \frac{\sum \lambda_j T_{cmj}}{\sum \lambda_j} \tag{5.1}$$

where
T_{mcm} = mean corrective maintenance time,
T_{cmj} = corrective maintenance time of the jth equipment/system element,
λ_j = failure rate of the jth equipment/system element.

Past experience indicates that probability distributions of corrective maintenance times follow exponential, normal, and lognormal. For example, in the case of electronic equipment with a good built-in test capability and a rapid remove and replace maintenance concept, often exponential distribution is assumed. In the case of mechanical or electro-mechanical hardware, usually with a remove and replace maintenance concept, the normal distribution is often applicable. Normally, the lognormal distribution is applicable to electronic equipment that does not possess built-in test capability.

MEDIAN ACTIVE CORRECTIVE MAINTENANCE TIME

This normally provides the best average location of the sample data and is the 50th percentile of all values of repair time. It may be said that median corrective maintenance time is a measure of the time within which 50% of all corrective maintenance can be accomplished. The computation of this measure depends on the distribution representing corrective maintenance times. Consequently, the median of the lognormally distributed corrective maintenance times is given by[6]

$$T_{med} = \text{antilog}\left(\frac{\Sigma \lambda_j \log T_{cmj}}{\Sigma \lambda_j}\right) \tag{5.2}$$

where
T_{med} = median active corrective maintenance time.

MAXIMUM ACTIVE CORRECTIVE MAINTENANCE TIME

This measures the time needed to accomplish all potential corrective maintenance actions up to a given percentage, frequently the 90th or 95th percentiles. For example, in the case of 90th percentile, the maximum corrective maintenance time is the time within which 90% of all maintenance actions can be accomplished. The distribution of corrective maintenance times dictates the calculation of the maximum corrective maintenance time. In the case of lognormally distributed corrective maintenance times, the maximum active corrective maintenance time is given by:[6]

$$T_{cmax} = \text{antilog}(T_{mn} + z\sigma_{cm}) \tag{5.3}$$

where
T_{cmax} = maximum active corrective maintenance time,
T_{mn} = mean of the logarithms of T_{cmj},
σ_{cm} = standard deviation of the logarithms of the sample corrective maintenance times,
z = standard deviation value corresponding to the percentile value specified for T_{cmax}.

The value of σ_{cm} can be calculated by using the following equation:

$$\sigma_{cm} = \left[\frac{\sum_{j=1}^{M}(\log T_{cmj})^2 - \left(\sum_{j=1}^{M}\log T_{cmj}\right)^2 / M}{M-1}\right]^{1/2} \tag{5.4}$$

where
M = total number of corrective maintenance times.

CORRECTIVE MAINTENANCE
MATHEMATICAL MODELS

Over the years a vast amount of literature has been published that directly or indirectly concerns corrective maintenance. This section presents a number of mathematical models taken from the published literature. These models take into consideration the item failure and corrective maintenance rates, and can be used to predict item/system

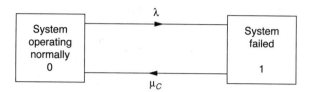

FIGURE 5.3 System transition diagram.

probability of being in failed state (i.e., undergoing repair), availability, mean time to failure, and so on.

MODEL I

This mathematical model represents a system that can either be in up (operating) or down (failed) state.[8] Corrective maintenance is performed on the failed system to put it back into its operating state. The system state space diagram is shown in Fig. 5.3.

Equations for the model are subject to the following assumptions:

- Failure and corrective maintenance rates are constant.
- The repaired system is as good as new.
- System failures are statistically independent.

The following symbols are used to develop equations for the model:

i = the ith system state, $i=0$ (system operating normally), $i=1$ (system failed);
$P_i(t)$ = probability that the system is in state i at time t;
λ = system failure rate;
μ_C = system corrective maintenance rate.

Using the Markov approach presented in Chapter 12, we write the following two equations for the Fig. 5.3 diagram:[8]

$$\frac{dP_0(t)}{dt} + \lambda P_0(t) = \mu_C P_1(t) \tag{5.5}$$

$$\frac{dP_1(t)}{dt} + \mu_C P_1(t) = \lambda P_0(t) \tag{5.6}$$

At time $t = 0$, $P_0(0) = 1$ and $P_1(t) = 0$.

By solving Eqs. (5.5) and (5.6), we get

$$P_0(t) = \frac{\mu_C}{\lambda + \mu_C} + \frac{\lambda}{\lambda + \mu_C} e^{-(\lambda + \mu_C)t} \tag{5.7}$$

$$P_1(t) = \frac{\mu}{\lambda + \mu_C} - \frac{\lambda}{\lambda + \mu_C} e^{-(\lambda + \mu_C)t} \tag{5.8}$$

The system availability is given by

$$A_S(t) = P_0(t) = \frac{\mu_C}{\lambda + \mu_C} + \frac{\lambda}{\lambda + \mu_C} e^{-(\lambda + \mu_C)t} \tag{5.9}$$

where
$A_S(t)$ = system availability at time t.

As t becomes very large, Eq. (5.9) reduces to

$$A_S = \frac{\mu_C}{\lambda + \mu_C} \tag{5.10}$$

where
A_S = system steady state availability.

Since $\lambda = 1/\text{MTTF}$ and $\mu_C = 1/T_{\text{mcm}}$, Eq. (5.10) becomes

$$A_S = \frac{\text{MTTF}}{T_{\text{mcm}} + \text{MTTF}} \tag{5.11}$$

where
MTTF = mean time to failure.

Example 5.1

Assume the MTTF of a piece of equipment is 3000 h and its mean corrective maintenance time is 5 h. Calculate the equipment steady-state availability, if the equipment failure and corrective maintenance times are exponentially distributed.

Substituting the given values into Eq. (5.11) yields

$$A_S = \frac{3000}{5 + 3000} = 0.9983$$

There is 99.83% chance that the equipment will be available for service.

MODEL II

This mathematical model represents a system that can either be operating normally or failed in two mutually exclusive failure modes (i.e., failure modes I and II). A typical example of this type of system or device is a fluid flow valve (i.e., open and close failure modes). Corrective maintenance is performed from either failure mode of the system to put it back into its operational state.[9-10]

The system transition diagram is shown in Fig. 5.4. The following assumptions are associated with this model:

- The system can fail in two mutually exclusive failure modes.
- The repaired system is as good as new.
- All system failures are statistically independent.
- Failure and corrective maintenance rates are constant.

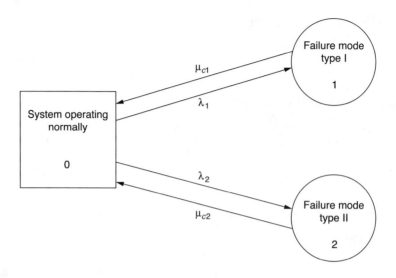

FIGURE 5.4 System transition diagram.

The following symbols are associated with the model:

i = the ith system state, $i = 0$ (system operating normally), $i = 1$ (system failed in failure mode type I), $i = 2$ (system failed in failure mode type II),

$P_i(t)$ = probability that the system is in state i at time t, for $i = 0, 1, 2$,

λ_i = system failure rate from state 0 to state i, for $i = 1, 2$,

μ_{Ci} = system corrective maintenance rate from state i to state 0, for $i = 1, 2$.

For Model I, we write the following equations for the Fig. 5.4 diagram:

$$\frac{dP_0(t)}{dt} + (\lambda_1 + \lambda_2)P_0(t) = \mu_{C1}P_1(t) + \mu_{C2}P_2(t) \tag{5.12}$$

$$\frac{dP_1(t)}{dt} + \mu_{C1}P_1(t) = \lambda_1 P_0(t) \tag{5.13}$$

$$\frac{dP_2(t)}{dt} + \mu_{C2}P_2(t) = \lambda_2 P_0(t) \tag{5.14}$$

At time $t = 0$, $P_0(0) = 1$ and $P_1(0) = P_2(0) = 0$.

By solving Eqs. (5.12)–(5.14), we obtain

$$P_0(t) = \frac{\mu_{C1}\mu_{C2}}{m_1 m_2} + \left[\frac{(m_1 + \mu_{C1})(m_1 + \mu_{C2})}{m_1(m_1 - m_2)}\right]e^{m_1 t} - \left[\frac{(m_2 + \mu_{C1})(m_2 + \mu_{C2})}{m_2(m_1 - m_2)}\right]e^{m_2 t}$$

$$\tag{5.15}$$

$$P_1(t) = \frac{\lambda_1 \mu_{C2}}{m_1 m_2} + \left[\frac{\lambda_1 m_1 + \lambda_1 \mu_{C2}}{m_1(m_1 - m_2)}\right]e^{m_1 t} - \left[\frac{(\mu_{C1} + m_2)\lambda_1}{m_2(m_1 - m_2)}\right]e^{m_2 t} \quad (5.16)$$

$$P_2(t) = \frac{\lambda_2 \mu_{C1}}{m_1 m_2} + \left[\frac{\lambda_2 m_1 + \lambda_2 \mu_{C1}}{m_1(m_1 - m_2)}\right]e^{m_1 t} - \left[\frac{(\mu_{C1} + m_2)\lambda_2}{m_2(m_1 - m_2)}\right]e^{m_2 t} \quad (5.17)$$

where

$$m_1, m_2 = \frac{-A \pm (A^2 - 4B)^{1/2}}{2} \quad (5.18)$$

$$A = \mu_{C1} + \mu_{C2} + \lambda_1 + \lambda_2 \quad (5.19)$$

$$B = \mu_{C1}\mu_{C2} + \lambda_1 \mu_{C2} + \lambda_2 \mu_{C1} \quad (5.20)$$

$$m_1 m_2 = \mu_{C1}\mu_{C2} + \lambda_1 \mu_{C2} + \lambda_2 \mu_{C1} \quad (5.21)$$

$$m_1 + m_2 = -(\mu_{C1} + \mu_{C2} + \lambda_1 + \lambda_2) \quad (5.22)$$

The system availability, $A_S(t)$, is given by

$$A_S(t) = P_0(t) \quad (5.23)$$

As time t becomes very large, from Eqs. (5.15) and (5.23), we get the following expression for the system steady state availability:

$$A_S = \frac{\mu_{C1}\mu_{C2}}{m_1 m_2} = \frac{\mu_{C1}\mu_{C2}}{\mu_{C1}\mu_{C2} + \lambda_1 \mu_{C2} + \lambda_2 \mu_{C1}} \quad (5.24)$$

Example 5.2

An engineering system can fail in two mutually exclusive failure modes. Failure modes I and II constant failure rates are $\lambda_1 = 0.002$ failures per hour and $\lambda_2 = 0.005$ failures per hour, respectively. The constant corrective maintenance rates from failure modes I and II are $\mu_{C1} = 0.006$ repairs per hour and $\mu_{C2} = 0.009$ repairs per hour, respectively. Calculate the system steady state availability.

Inserting the specified values into Eq. (5.24) yields

$$A_S = \frac{0.006 \times 0.009}{(0.006 \times 0.009) + (0.002 \times 0.009) + (0.005 \times 0.006)}$$

$$= 0.5294$$

Thus, the system steady state availability is 0.5294. There is approximately a 53% chance that the system will be available for service when needed.

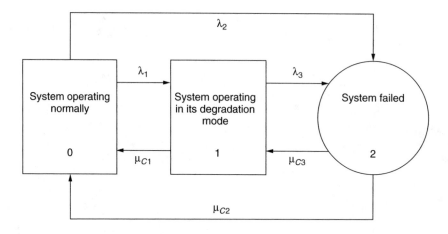

FIGURE 5.5 System transition diagram.

MODEL III

This mathematical model represents a system that can either be operating normally, operating in degradation mode, or failed completely. An example of this type of system could be a power generator, i.e., producing electricity at full capacity, derated capacity, or not at all. Corrective maintenance is initiated from degradation and completely failed modes of the system to repair failed parts.[10] The system state space diagram is shown in Fig. 5.5. The model is subject to the following assumptions:

- System complete failure, partial failure, and corrective maintenance rates are constant.
- The operating system can either fail fully or partially. The partially operating system can stop operating altogether.
- All system failures are statistically independent.
- The repaired system is as good as new.

The following symbols are associated with the model:

i = the ith system state, $i = 0$ (system operating normally), $i = 1$ (system operating in its degradation mode), $i = 2$ (system failed),

$P_i(t)$ = probability that the system is in state i at time t, for $i = 0, 1, 2$,

λ_i = system failure rate, $i = 1$ (from state 0 to state 1), $i = 2$ (from state 0 to state 2), $i = 3$ (from state 1 to state 2),

μ_{Ci} = system corrective maintenance rate, $i = 1$ (from state 1 to state 0), $i = 2$ (from state 2 to state 0), $i = 3$ (from state 2 to state 1).

For Models I and II, we write the following equations for the Fig. 5.5 diagram:

$$\frac{dP_0(t)}{dt} + (\lambda_1 + \lambda_2)P_0(t) = \mu_{C1}P_1(t) + \mu_{C2}P_2(t) \qquad (5.25)$$

$$\frac{dP_1(t)}{dt} + (\mu_{C1} + \lambda_3)P_1(t) = \mu_{C3}P_2(t) + \lambda_1 P_0(t) \tag{5.26}$$

$$\frac{dP_2(t)}{dt} + (\mu_{C2} + \mu_{C3})P_2(t) = \lambda_3 P_1(t) + \lambda_2 P_0(t) \tag{5.27}$$

At time $t = 0$, $P_0(0) = 1$ and $P_1(0) = P_2(0) = 0$. By solving Eqs. (5.25)–(5.27), we get

$$P_0(t) = \frac{(\mu_{C1}\mu_{C2} + \lambda_3\mu_{C2} + \mu_{C1}\mu_{C3})}{K_1 K_2}$$

$$+ \left[\frac{\mu_{C1}K_1 + \mu_{C2}K_2 + \mu_{C3}K_1 + K_1\lambda_3 + K_1^2 + \mu_{C1}\mu_{C2} + \lambda_3\mu_{C2} + \mu_{C1}\mu_{C3}}{Y}\right]e^{K_1 t}$$

$$+ \left\{1 - \left(\frac{\mu_{C1}\mu_{C2} + \lambda_3\mu_{C2} + \mu_{C1}\mu_{C3}}{K_1 K_2}\right)\right.$$

$$\left. - \left[\frac{\mu_{C1}K_1 + \mu_{C2}K_1 + \mu_{C3}K_1 + K_1\lambda_3 + K_2^2 + \mu_{C1}\mu_{C2} + \lambda_3\mu_{C2} + \mu_{C1}\mu_{C3}}{Y}\right]\right\}e^{K_2 t}$$

$$\tag{5.28}$$

where
$$Y = K_1(K_1 - K_2).$$

$$P_1(t) = \left(\frac{\lambda_1\mu_{C2} + \lambda_1\mu_{C3} + \lambda_2\mu_{C3}}{K_1 K_2}\right) + \left[\frac{K_1\lambda_1 + \lambda_1\mu_{C2} + \lambda_1\mu_{C3} + \lambda_2\mu_{C3}}{Y}\right]e^{K_1 t}$$

$$- \left[\frac{\lambda_1\mu_{C2} + \lambda_1\mu_{C3} + \lambda_2\mu_{C3}}{K_1 K_2} + \frac{K_1\lambda_1 + \lambda_1\mu_{C2} + \lambda_1\mu_{C3} + \lambda_2\mu_{C3}}{Y}\right]e^{K_2 t} \tag{5.29}$$

$$P_2(t) = \left(\frac{\lambda_1\lambda_3 + \mu_{C1}\lambda_2 + \lambda_2\lambda_3}{K_1 K_2}\right) + \left[\frac{K_1\lambda_2 + \lambda_1\lambda_3 + \lambda_2\mu_{C1} + \lambda_2\lambda_3}{Y}\right]e^{K_1 t}$$

$$- \left[\frac{\lambda_1\lambda_3 + \mu_{C1}\lambda_2 + \lambda_2\lambda_3}{K_1 K_2} + \frac{\lambda_2 K_1 + \lambda_1\lambda_3 + \mu_{C1}\lambda_2 + \lambda_2\lambda_3}{Y}\right]e^{K_2 t} \tag{5.30}$$

where

$$K_1, K_2 = \frac{-D \pm (D^2 - 4F)^{1/2}}{2} \tag{5.31}$$

$$D = \mu_{C1} + \mu_{C2} + \mu_{C3} + \lambda_1 + \lambda_2 + \lambda_3 \tag{5.32}$$

$$F = K_1 K_2 = \mu_{C1}\mu_{C2} + \lambda_3\mu_{C2} + \mu_{C1}\mu_{C3} + \mu_{C2}\lambda_1 + \lambda_1\mu_{C3}$$
$$+ \lambda_1\lambda_3 + \mu_{C1}\lambda_2 + \lambda_2\mu_{C3} + \lambda_2\lambda_3 \tag{5.33}$$

The system full/partial availability, $A_{Sf/p}(t)$, is given by

$$A_{Sf/p}(t) = P_0(t) + P_1(t) \tag{5.34}$$

As t becomes large, Eq. (5.34) reduces to

$$A_{Sf/p} = \frac{\mu_{C1}\mu_{C2} + \lambda_3\mu_{C2} + \mu_{C1}\mu_{C3} + \lambda_1\mu_{C2} + \lambda_1\mu_{C3} + \lambda_2\mu_{C3}}{K_1 K_2} \tag{5.35}$$

where

$A_{Sf/p}$ = system full/partial steady-state availability.

Similarly, the system full steady-state availability is

$$A_{Sf} = P_0 = \frac{\mu_{C1}\mu_{C2} + \lambda_3\mu_{C2} + \mu_{C1}\mu_{C3}}{K_1 K_2} \tag{5.36}$$

Example 5.3

Assume that in Eq. (5.36), we have $\lambda_1 = 0.002$ failures per hour, $\lambda_2 = 0.003$ failures per hour, $\lambda_3 = 0.001$ failures per hour, $\mu_{C1} = 0.006$ repairs per hour, $\mu_{C2} = 0.004$ repairs per hour, and $\mu_{C3} = 0.008$ repairs per hour. Calculate the value of the system full steady-state availability.

Inserting the specified data values into Eq. (5.33) yields

$$\begin{aligned}
K_1 K_2 &= (0.006 \times 0.004) + (0.001 \times 0.004) + (0.006 \times 0.008) \\
&\quad + (0.004 \times 0.002) + (0.002 \times 0.008) + (0.002 \times 0.001) \\
&\quad + (0.006 \times 0.003) + (0.003 \times 0.008) + (0.003 \times 0.001) \\
&\approx 0.0001
\end{aligned}$$

Using the above calculated value and the given data in Eq. (5.36) we get $A_{Sf} = 0.5170$. There is approximately 52% chance that the system will be available for full service.

Model IV

This mathematical model represents a two identical-unit redundant (parallel) system. At least one unit must operate normally for system success. Corrective maintenance to put it back into its operating state begins as soon as any one of the units fails.[11] The system state space diagram is shown in Fig. 5.6.

The following assumptions are associated with the model:

- The system is composed of two independent and identical units.
- The repaired unit is as good as new.
- No corrective maintenance is performed on the failed system (i.e., when both units fail).
- Unit failure and corrective maintenance rates are constant.

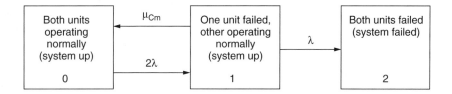

FIGURE 5.6 Two identical unit redundant system transition diagram.

The following symbols pertain to the model:

i = the ith system state, $i = 0$ (both units operating normally), $i = 1$ (one unit failed, other operating), $i = 2$ (both units failed),

$P_i(t)$ = probability that the system is in state i at time t, for $i = 0, 1, 2,$

λ = unit failure rate,

μ_{Cm} = unit corrective maintenance rate.

For Models I, II, and III we write the following equations for the Fig. 5.6 transition diagram:

$$\frac{dP_0(t)}{dt} + 2\lambda P_0(t) = \mu_{Cm}P_1(t) \tag{5.37}$$

$$\frac{dP_1(t)}{dt} + (\mu_{Cm} + \lambda)P_1(t) = 2\lambda P_0(t) \tag{5.38}$$

$$\frac{dP_2(t)}{dt} = \lambda P_1(t) \tag{5.39}$$

At time $t = 0$, $P_0(0) = 1$, and $P_1(0) = P_2(0) = 0$. Solving Eqs. (5.37)–(5.39), we get

$$P_0(t) = \left[\frac{\lambda + \mu + C_1}{C_1 - C_2}\right]e^{C_1 t} - \left[\frac{\lambda + \mu + C_1}{C_1 - C_2}\right]e^{C_2 t} \tag{5.40}$$

$$P_1(t) = \left[\frac{2\lambda}{C_1 - C_2}\right]e^{C_1 t} - \left[\frac{2\lambda}{C_1 - C_2}\right]e^{C_2 t} \tag{5.41}$$

$$P_2(t) = 1 + \left[\frac{C_2}{C_1 - C_2}\right]e^{C_1 t} - \left[\frac{C_1}{C_1 - C_2}\right]e^{C_2 t} \tag{5.42}$$

where

$$C_1, C_2 = \frac{[-(3\lambda + \mu) \pm (3\lambda + \mu)^2 - 8\lambda^2]^{1/2}}{2} \tag{5.43}$$

$$C_1 C_2 = 2\lambda^2 \tag{5.44}$$

$$C_1 + C_2 = -(3\lambda + \mu) \tag{5.45}$$

The system reliability is given by

$$R_S(t) = P_0(t) + P_1(t) \tag{5.46}$$

where
$R_S(t)$ = redundant system reliability at time t.

The system mean time to failure (MTTF_S) is given by

$$\text{MTTF}_S = \int_0^\infty R_S(t)dt = \frac{3\lambda + \mu_{\text{Cm}}}{2\lambda^2} \tag{5.47}$$

Since $\lambda = 1/\text{MTTF}_u$ and $\mu_{\text{Cm}} = 1/\text{MCMT}$, Eq. (5.47) becomes

$$\text{MTTF}_S = \frac{\text{MTTF}_u}{2\text{MCMT}}(3\text{MCMT} + \text{MTTF}_u) \tag{5.48}$$

where
MTTF_u = unit mean time to failure,
MCMT = unit mean corrective maintenance time.

Example 5.4

A system is composed of two independent and identical units in parallel. A failed unit is repaired immediately but the failed system is never repaired. The unit times to failure and corrective maintenance times are exponentially distributed. The unit mean time to failure and mean corrective maintenance time are 150 h and 5 h, respectively. Calculate the system mean time to failure with and without the performance of corrective maintenance.

Inserting the given values into Eq. (5.48), we get

$$\text{MTTF}_S = \frac{150}{2 \times 5}(3 \times 5 + 150) = 2475 \text{ h}$$

Setting $\mu_{\text{Cm}} = 0$ and substituting the given value into Eq. (5.47) yields

$$\text{MTTF}_S = \frac{3}{2\lambda} = 225 \text{ h}$$

This means introduction of corrective maintenance helped increase system mean time to failure from 225 h to 2475 h.

APPROXIMATE EFFECTIVE FAILURE RATE EQUATIONS FOR REDUNDANT SYSTEMS WITH CORRECTIVE MAINTENANCE

This section presents approximate effective failure rate equations for two types of redundant systems. The effective failure rate is the reciprocal of the item/system mean time to failure.

EFFECTIVE FAILURE RATE OF SYSTEM TYPE I

The system type I is assumed to contain m number of independent and identical active units in parallel and in which k units are allowed to fail without system failure. The corrective maintenance begins as soon as a unit fails. The failed system is never repaired. The unit failure and corrective maintenance rates are constant. Thus, an approximate effective failure rate of a $(m - K)$-out-of-m system can be calculated using the following equation:[12]

$$\lambda_{(m-K)/m} = \frac{m!\lambda^{K+1}}{(m - K - 1)!\mu^{K}} \qquad (5.49)$$

where

$\lambda_{(m-K)/m}$ = system approximate effective failure rate. In this system at least $(m - K)$ units must work normally for the system success,

λ = unit failure rate,

μ = unit corrective maintenance rate.

Example 5.5

A system is composed of three independent and identical units in parallel and at least two units must operate normally for the system success. The unit failure rate is 0.0001 failures per hour. It takes an average of 2 h to repair (exponential distribution) a failed unit to an active state. Calculate the system approximate effective failure rate if the failed system is never repaired.

$$\begin{aligned}
\lambda_{(3-1)/3} &= \frac{3!\lambda^{2}}{1!\mu} = \frac{6\lambda^{2}}{\mu} \\
&= \frac{6(0.0001)^{2}}{0.5} \\
&= 1.2 \times 10^{-7} \text{ failures per hour}
\end{aligned}$$

The system effective failure rate is 1.2×10^{-7} failures per hour.

EFFECTIVE FAILURE RATE OF SYSTEM TYPE II

System type II is composed of two independent and nonidentical units in parallel. Corrective maintenance begins as soon as either unit fails. The failed system is never repaired. Unit failure and corrective maintenance rates are constant.

An approximate formula to obtain system effective failure rate is as follows:[12]

$$\lambda_{se} = \frac{\lambda_1 \lambda_2 [(\mu_1 + \mu_2) + (\lambda_1 + \lambda_2)]}{\mu_1 \mu_2 + (\mu_1 + \mu_2)(\lambda_1 + \lambda_2)} \tag{5.50}$$

where
λ_{se} = two unit parallel system effective failure rate,
λ_i = unit i failure rate, for $i = 1, 2$,
μ_i = unit i corrective maintenance rate, for $i = 1, 2$.

Example 5.6

A system is composed of two independent and nonidentical units in parallel. Unit 1 failure and corrective maintenance rates are 0.004 failures per hour and 0.005 repairs per hour, respectively. Similarly, the unit 2 failure and corrective maintenance rates are 0.002 failures per hour and 0.003 repairs per hour, respectively. The failed system is never repaired. Calculate the system approximate effective failure rate.

Substituting the given data into Eq. (5.50), we get

$$\lambda_{se} = \frac{(0.004 \times 0.002)[(0.005 + 0.003) + (0.004 + 0.002)]}{(0.005 \times 0.003) + (0.005 + 0.003)(0.004 + 0.002)}$$

$$= 0.0018 \text{ failures per hour}$$

The system effective failure rate is 0.0018 failures per hour.

PROBLEMS

1. Define corrective maintenance.
2. Describe the following types of corrective maintenance:
   • Overhaul
   • Rebuild
   • Servicing
3. Discuss sequential steps associated with corrective maintenance.
4. Define main components of active repair time.
5. Discuss at least four strategies for reducing the system-level corrective maintenance time.
6. Define median corrective maintenance time.
7. Assume that exponential mean time to failure and mean corrective maintenance time of a system are 2500 h and 4 h, respectively. Calculate the system steady-state availability.

8. A system can fail in two mutually exclusive failure modes. Failure mode I constant failure and corrective maintenance rates are 0.005 failures per hour and 0.02 repairs per hour, respectively. Similarly, failure mode II constant failure and corrective maintenance rates are 0.001 failures per hour and 0.03 repairs per hour, respectively. Calculate the system steady-state availability.

9. A system is composed of two independent and identical units in parallel. Although a failed unit is repaired immediately, the failed system is never repaired. The unit times to failure and corrective maintenance times are exponentially distributed. Thus, the unit mean time to failure and mean corrective maintenance time are 200 h and 2 h, respectively. Calculate the system mean time to failure.

10. Assume that a system is composed of two independent and identical units in parallel and at least one unit must operate normally for system success. The unit failure and repair rates are 0.002 failures per hour and 0.01 repairs per hour, respectively. The failed system is never repaired. Calculate the value of the system approximate effective failure rate.

REFERENCES

1. AMCP 706-132, *Engineering Design Handbook: Maintenance Engineering Techniques,* Department of Defense, Washington, D.C., 1975.

2. Omdahl, T.P., *Reliability, Availability, and Maintainability (RAM) Dictionary,* ASQC Quality Press, Milwaukee, Wisconsin, 1988.

3. McKenna, T. and Oliverson, R., *Glossary of Reliability and Maintenance Terms,* Gulf Publishing Company, Houston, Texas, 1997.

4. MICOM 750-8, *Maintenance of Supplies and Equipment,* Department of Defense, Washington, D.C., March 1972.

5. NAVORD OD 39223, *Maintainability Engineering Handbook,* Department of Defense, Washington, D.C., June 1969.

6. Blanchard, B.S., Verma, D., and Peterson, E.L., *Maintainability,* John Wiley & Sons, New York, 1995.

7. AMCP-766-133, *Engineering Design Handbook: Maintainability Engineering Theory and Practice,* Department of Defense, Washington, D.C., 1976.

8. Dhillon, B.S., *Design Reliability: Fundamentals and Application,* CRC Press, Boca Raton, Florida, 1999.

9. Dhillon, B.S. and Singh, C., *Engineering Reliability: New Techniques and Applications,* John Wiley & Sons, New York, 1981.

10. Dhillon, B.S., *Reliability Engineering in Systems Design and Operation,* Van Nostrand Reinhold Co., New York, 1983.

11. Shooman, M.L., *Probabilistic Reliability: An Engineering Approach,* McGraw-Hill, New York, 1968.

12. *RADC Reliability Engineer's Toolkit,* prepared by the Systems Reliability and Engineering Division, Rome Air Development Center, Griffiss Air Force Base, Rome, New York, July 1988.

6 Reliability Centered Maintenance

INTRODUCTION

Reliability centered maintenance (RCM) is a systematic process used to determine what has to be accomplished to ensure that any physical facility is able to continuously meet its designed functions in its current operating context.[1] RCM leads to a maintenance program that focuses preventive maintenance (PM) on specific failure modes likely to occur. Any organization can benefit from RCM if its breakdowns account for more than 20 to 25% of the total maintenance workload.[2]

With the arrival of the Boeing 747, a wide-body aircraft, U.S. airlines realized that their maintenance activity would require considerable change due to a large increase in scheduled maintenance costs. In 1968, airline operators jointly organized a study group to develop methodology for resolving the problem. The group was called Maintenance Steering Group No. 1 (MSG1).[3] The resulting documents, MSG1,[4] MSG2,[5] and MSG3,[6] appeared in 1968, 1970, and 1980, respectively.[7]

The term "reliability centered maintenance" appeared for the first time as the title of a report on the processes used by the civil aviation industry to prepare maintenance programs for aircraft.[8,9] The report, prepared by United Airlines, was commissioned by the U.S. Department of Defense in 1974.[10] The history of RCM is described in detail in References 8 through 13. This chapter presents important aspects of RCM.

RCM GOALS AND PRINCIPLES

Some of the important goals of RCM are as follows:[3]

- To develop design-associated priorities that can facilitate PM.
- To gather information useful for improving the design of items with proven unsatisfactory, inherent reliability.
- To develop PM-related tasks that can reinstate reliability and safety to their inherent levels in the event of equipment or system deterioration.
- To achieve the above goals when the total cost is minimal.

Many principles of RCM are discussed below:[10]

- *RCM is system/equipment focused.* RCM is concerned more with maintaining system function as opposed to maintaining individual component function.

- *Safety and economics drive RCM.* Safety is of paramount importance, thus it must be ensured at any cost and then cost effectiveness becomes the criterion.
- *RCM is function-oriented.* RCM plays an instrumental role in preserving system/equipment function, not just operability for its own sake.
- *Design limitations are acknowledged by RCM.* The goal of RCM is to maintain the inherent reliability of the equipment/system design and at the same time recognize that changes in inherent reliability can only be made through design rather than maintenance. Maintenance at the best of times can only achieve and maintain a level of designed reliability.
- *RCM is reliability-centered.* RCM is not overly concerned with simple failure rate, but it places importance on the relationship between operating age and failures experienced. RCM treats failure statistics in an actuarial fashion.
- *An unsatisfactory condition is defined as a failure by RCM.* A failure could be either a loss of acceptable quality or a loss of function.
- *RCM is a living system.* RCM collects information from the results achieved and feeds it back to improve design and future maintenance.
- *Three types of maintenance tasks along with run-to-failure are acknowledged by RCM.* These tasks are defined as failure-finding, time-directed, and condition-directed. The purpose of the failure-finding tasks is to discover hidden functions that have failed without providing any indication of pending failure. Time-directed tasks are scheduled as considered necessary. Condition-directed tasks are conducted as the conditions indicate for their need. Run-to-failure is a conscious decision in RCM.
- *RCM tasks must be effective.* The tasks must be cost-effective and technically sound.
- *RCM uses a logic tree to screen maintenance tasks.* This provides consistency in the maintenance of all types of equipment.
- *RCM tasks must be applicable.* Tasks must reduce the occurrence of failures or ameliorate secondary damage resulting from failure.

RCM PROCESS AND ASSOCIATED QUESTIONS

The RCM process is applied to determine particular maintenance tasks to be performed, as well as to influence item reliability and maintainability during design. Initially the RCM process is applied during the design and development phase and then reapplied, as appropriate, during the operational phase to sustain an effective maintenance program based on experience in the field. Any RCM process should ensure that all of the following questions are answered effectively as per their sequence:[14]

- What are the functions and associated expected levels of the facility performance in its current operating context?
- How might it fail to meet its assigned functions?
- What are the reasons for each functional failure or failure mode?

- What are the effects of each failure?
- How does each failure matter?
- What remedial measures should be taken to prevent or predict each failure?
- What measures should be taken in the event of not finding a suitable proactive task?

The basic RCM process is composed of the following steps:[15]

1. *Identify important items with respect to maintenance.* Usually, maintenance-important items are identified using techniques such as failure, mode, effects, and criticality analysis (FMECA) and fault tree analysis (FTA).
2. *Obtain appropriate failure data.* In determining occurrence probabilities and assessing criticality, the availability of data on part failure rate, operator error probability, and inspection efficiency is essential. These types of data come from field experience, generic failure databanks, etc. Many sources for obtaining failure data are listed in Reference 16.
3. *Develop fault tree analysis data.* Probabilities of occurrence of fault events— basic, intermediate, and top events—are calculated as per combinatorial properties of the logic elements in the fault tree.
4. *Apply decision logic to critical failure modes.* The decision logic is designed to lead, by asking standard assessment questions, to the most desirable preventive maintenance task combinations. The same logic is applied to each crucial mode of failure of each maintenance-important item.
5. *Classify maintenance requirements.* Maintenance requirements are categorized into three classifications: on-condition maintenance requirements, condition-monitoring maintenance requirements, and hard-time maintenance requirements.
6. *Implement RCM decisions.* Task frequencies and intervals are set/enacted as part of the overall maintenance strategy or plan.
7. *Apply sustaining-engineering on the basis of field experience.* Once the system/equipment start operating, the real-life data begin to accumulate. At that time, one of the most urgent steps is to re-evaluate all RCM-associated default decisions.

RCM COMPONENTS

The four major components of RCM are shown in Fig. 6.1.[11,17] These are: reactive maintenance, preventive maintenance, predictive testing and inspection, and proactive maintenance. Each component is described below.

REACTIVE MAINTENANCE

This type of maintenance is also known as breakdown, fix-when-fail, run-to-failure, or repair maintenance. When using this maintenance approach, equipment repair, maintenance, or replacement takes place only when deterioration in the condition

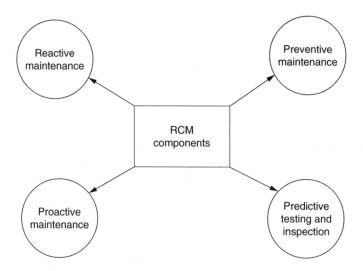

FIGURE 6.1 Components of RCM.

of an item/equipment results in a functional failure. In this type of maintenance, it is assumed there is an equally likely chance for the occurrence of a failure in any part, component, or system. When reactive maintenance is practiced solely, a high replacement of part inventories, poor use of maintenance effort, and high percentage of unplanned maintenance activities are typical. Furthermore, an entirely reactive maintenance program overlooks opportunities to influence equipment/item survivability.

Reactive maintenance can be practiced effectively only if it is carried out as a conscious decision, based on the conclusions of an RCM analysis that compares risk and cost of failure with the cost of maintenance needed to mitigate that risk and failure cost. A criteria for determining the priority of replacing or repairing the failed item/equipment in the reactive maintenance program is presented in Table 6.1.[11]

PREVENTIVE MAINTENANCE

Preventive maintenance (PM), also called time-driven or interval-based maintenance, is performed without regard to equipment condition. It consists of periodically scheduled inspection, parts replacement, repair of components/items, adjustments, calibration, lubrication, and cleaning. PM schedules regular inspection and maintenance at set intervals to reduce failures for susceptible equipment. It is important to note that, depending on the predefined intervals, practicing PM can lead to a significant increase in inspections and routine maintenance. On the other hand, it can help reduce the frequency and severity of unplanned failures. Preventive maintenance can be costly and ineffective if it is the only type of maintenance practiced.

TABLE 6.1
Reactive Maintenance Priority Classifications

Priority Description	Priority Level	Criteria Based on System/Equipment Failure Consequences
Emergency	I	Serious and an immediate impact on mission
		Safety of life/property is under threat
Urgent	II	Serious and an impending impact on mission
		Continuity of facility operation is threatened
Priority	III	Significant and adverse effect on project is imminent
		Degradation in quality of mission support
Routine	IV	Insignificant impact on mission
		Existence of redundancy
Discretionary	V	Resources are available
		Impact on mission is negligible
Deferred	VI	Unavailability of resources
		Negligible impact on mission

PM Task and Monitoring Periodicity Determination

Even though there are many ways to determine the correct periodicity of PM tasks, none are valid until the in-service age-reliability characteristics of the item affected by the desired tasks are known. Usually, this type of information is not available, but it must be obtained for new items/equipment. Past experience shows that predictive testing and inspection (PTI) techniques are useful in determining item/equipment condition vs. age.

Often, the parameter mean time between failures (MTBF) is used as the basis for determining the PM interval. This approach is considered wrong because it does not provide information about the effect of increasing age on item reliability. More specifically, the approach provides the average age at which failure occurs, but not the most likely age for the item under consideration. In the event of inadequate information on the effect of age on reliability, the most appropriate approach would be to monitor the item's condition.[11]

Item/Equipment Monitoring

The main objectives in monitoring item/equipment condition are to determine item/equipment condition and to establish a trend to forecast future item/equipment condition. The following approaches are useful for setting initial periodicity:

- *Failure anticipation from past experience:* In some cases, failure history of equipment and personal experience can provide, to a certain degree, an intuitive feel as when to expect a failure.

- *Failure distribution statistics:* The failure distribution and the probability of failure must be known when statistics are used to determine the basis for selecting periodicities.
- *Conservative approach:* The common practice in the industrial sector is to monitor the equipment monthly/weekly when good monitoring methods and adequate information are unavailable. Often, this leads to excessive monitoring. In situations when impending failure becomes apparent through the use of trending or other predictive analysis techniques, the monitoring interval can be shortened.

PREDICTIVE TESTING AND INSPECTION

Predictive testing and inspections (PTI) is sometimes called condition monitoring or predictive maintenance. To assess item/equipment condition, it uses performance data, nonintrusive testing techniques, and visual inspection. PTI replaces arbitrarily timed maintenance tasks with maintenance that is performed as warranted by the item/equipment condition. Analysis of item/equipment condition-monitoring data on a continuous basis is useful for planning and scheduling maintenance/repair in advance of catastrophic or functional failure.

The collected PTI data are used to determine the equipment condition and to highlight the precursors of failure in several ways, including pattern recognition, trend analysis, correlation of multiple technologies, data comparison, statistical process analysis, and tests against limits and ranges. PTI should not be the only type of maintenance practiced, because it does not lend itself to all types of items/equipment or possible modes of failure.

PROACTIVE MAINTENANCE

This type of maintenance helps improve maintenance through actions such as better design, workmanship, installation, scheduling, and maintenance procedures. The characteristics of proactive maintenance include practicing a continuous process of improvement, using feedback and communications to ensure that changes in design/procedures are efficiently made available to item designers/management, ensuring that nothing affecting maintenance occurs in total isolation, with the ultimate goal of correcting the concerned equipment forever, optimizing and tailoring maintenance methods and technologies to each application. It performs root-cause failure analysis and predictive analysis to enhance maintenance effectiveness, conducts periodic evaluation of the technical content and performance interval of maintenance tasks, integrates functions with support maintenance into maintenance program planning, and uses a life cycle view of maintenance and supporting functions.[11]

Figure 6.2 presents eight basic methods employed by proactive maintenance to extend item/equipment life. Some of these methods are described below.[11]

Reliability Engineering

Reliability engineering, in conjunction with other proactive maintenance approaches, involves the redesign, modification, or improvement of items/parts or their replacement

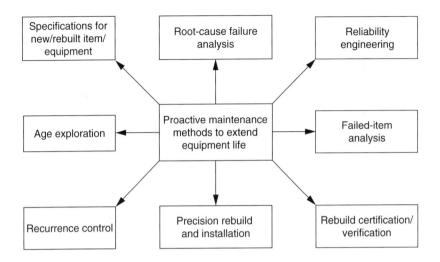

FIGURE 6.2 Basic techniques employed by proactive maintenance to extend equipment life.

with better item/parts. In some cases, a complete redesign of the item/part may be required. There are many techniques used in reliability engineering to perform reliability analysis of engineering items/systems. The two most widely used in the industrial sector are known as failure modes and effect analysis (FMEA) and fault tree analysis (FTA). The introductory aspects of reliability engineering are presented in Chapter 12.

Failed-Item Analysis

This involves visually inspecting failed items after removal to determine the reasons for failure. As the need arises, more detailed technical analysis is performed to find the real cause of a failure. For example, in the case of bearings, the root causes of their failures may relate to factors such as poor lubrication practices, excessive balance and alignment tolerances, improper installation, or poor storage and handling methods.

Past experience indicates that over 50% of all bearing problems are caused by improper installation or contamination. Usually, indicators of improper installation problems are evident on both internal and external surfaces of bearings and the indicators of contamination appear on the bearings' internal surfaces.

Root Cause Failure Analysis

Root cause failure analysis (RCFA) is concerned with proactively seeking the basic causes of facility/equipment failure. The main objectives of RCFA are to: determine the cause of a problem efficiently and economically, rectify the problem cause, not just its effect, instill a mentality of "fix forever," and provide data that can be useful in eradicating the problem.

Specifications for New/Rebuilt Item/Equipment

Here, the basic concern is with writing effective specifications, documenting problems, and testing the equipment of different vendors. At minimum, the specifications should include such items as vibration, balancing criteria, and alignment. The basis of this proactive approach is to document historical data so that involved professionals can effectively write verifiable purchasing and installation specifications for new/rebuilt equipment.

Age Exploration

Age exploration (AE) is an important factor in establishing an RCM program. It provides a mechanism to vary a maintenance program's key aspects to optimize the process. The AE approach examines the applicability of all maintenance tasks with respect to the following three factors:

1. *Technical content:* The task's technical contents are examined to ensure that all identified modes of failure are properly addressed, as well assuring that the existing maintenance tasks lead to the expected degree of reliability.
2. *Performance interval:* Adjustments are made continually to the task performance interval until the rate at which resistance to failure declines is effectively sensed or determined.
3. *Task grouping:* Tasks with similar periodicity are grouped for the purpose of improving the time spent on the job site and reducing outages.

Rebuild Certification/Verification

At the installation of new/rebuilt item/equipment, it is essential to verify that it is functioning effectively. Past experience indicates that it is a good practice to test the item/equipment against formal certification and verification standards to avoid unsatisfactory operating performance or early failure.

Recurrence Control

Recurrence control concerns the control of repetitive failures. Repetitive failures are defined as the recurring inability of an item to carry out the required function. The following situations fall under the category of repetitive failures:

- Repeated failure of a piece of equipment
- Repeated failure of items belonging to a system or subsystem
- Failures of the same/similar parts in various different equipment or systems

A process for performing an analysis of repetitive failures is presented in Fig. 6.3.

Precision Rebuild and Installation

To control life cycle costs and maximize reliability, the equipment under consideration requires proper installation. Often maintenance personnel and operators are faced with problems caused by poor equipment installation. Usually, two common rework items,

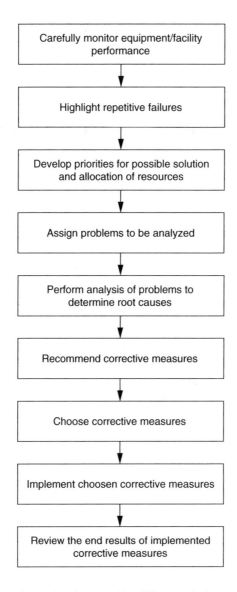

FIGURE 6.3 A process for performing repetitive failure analysis.

rotor balance and alignment, are unsatisfactorily performed or neglected during the initial installation phase. The effective application of precision standards can more than double equipment life.[11]

Past experience indicates that parasitic load, caused by imbalance and misalignment, is one of the leading causes of premature rolling element bearing failure. One important and cost-effective method for increasing bearing life and resultant equipment reliability is the precision balance of motor rotors, fans, and pump impellers.

Due to misalignment, the forces of vibration lead to gradual deterioration of seals, drive windings, couplings, and other rotating elements with close tolerances. A petrochemical industry survey indicates that the practice of precision alignment resulted in the average bearing life increasing by a factor of 8, reduction in maintenance costs by 7%, and a 12% increment in machinery availability.[11]

PREDICTIVE TESTING AND INSPECTION TECHNOLOGIES

Predictive testing and inspection (PTI) is an important component of the RCM. This section describes the PTI technologies in detail. These technologies may be described as a variety of approaches used to determine item/equipment condition for the purpose of estimating the most effective time to schedule maintenance. These technologies include intrusive and nonintrusive approaches in addition to using process parameters to assess overall condition of equipment.

Six PTI technologies/approaches are described below.[11,17]

1. *Vibration monitoring and analysis:* One of the most widely used PTI approaches, it is useful in assessing the condition of rotating equipment and structural stability in a system. The techniques of vibration monitoring and analysis include spectrum analysis, torsional vibration, waveform analysis, shock pulse analysis, and multichannel vibration analysis.

   Vibration monitoring effectiveness depends on such factors as analyst's ability, complexity of equipment, sensor mounting, resolution, and data collection methods. The vibration monitoring and analysis approach is applicable to items such as engines, shafts, motors, pumps, gearboxes, bearings, turbines, and compressors.

2. *Electrical condition monitoring:* This includes various technologies and approaches that provide a comprehensive system evaluation. By monitoring important electrical parameters it provides useful data to detect and rectify electrical related faults such as phase imbalance, insulation breakdown, and high resistance connections. Electrical faults are costly and present safety concerns because in systems they are seldom visible.

   Table 6.2[11] lists several electrical condition monitoring methods. These methods can monitor equipment such as electrical motors, electrical distribution cabling, generators, electrical distribution transformers, electrical distribution switchgear and controllers, and distribution systems. The specific electrical condition monitoring methods for these six types of equipment are presented in Table 6.3.

3. *Thermography:* Infrared thermography (IRT) may be defined as the application of infrared detection instruments for identifying pictures of temperature differences (thermogram). The test instruments used include noncontact, thermal measurement, line-of-sight, and imaging systems. The noncontact nature of the IRT technique makes it particularly attractive for identifying hot/cold spots in energized electrical equipment, large surface areas such as boilers and building walls, and so on.

TABLE 6.2
Electrical Condition Monitoring Methods

Method Name

Surge testing
Motor circuit analysis
Radio frequency monitoring
Infrared thermography
Motor current spectrum analysis
Airborne ultrasonics
Megohmmeter testing
Turns ratio
Transformer oil analysis
High potential testing
Time domain reflectometry
Starting current and time
Motor current readings
Power factor and harmonic distortion
Conductor complex impedance

TABLE 6.3
Electrical Condition Monitoring Techniques for Specific Equipment

Equipment	Applicable Monitoring Techniques
Generators	Radio frequency monitoring and megohmmeter testing
Electrical distribution transformers	Turns ratio, transformer oil analysis, power factor, and harmonic distortion
Electrical distribution switchgear and controllers	Airborne ultrasonics and visual inspection
Electrical motors	Motor current spectrum analysis, starting current, motor circuit analysis, megohmmeter testing, conductor complex impedance, surge testing, and high potential testing
Distribution system	Airborne ultrasonics, power factor, harmonic distortion, and high potential testing
Electrical distribution cabling	Airborne ultrasonics, high potential testing, megohmmeter testing, and time domain reflectometry

With respect to specific electrical equipment, the IRT approach can be used to identify degrading conditions in items such as switchgear, motor control centers, transformers, and substations. Similarly, in regard to specific mechanical equipment, the IRT technique can help identify blocked flow conditions in items such as condensers, pipes, heat exchangers, and transformer cooling radiators.

TABLE 6.4
Lubricating Oil and Hydraulic
Fluid Analysis Associated
Standard Analytical Tests

Test Name

Particle counting
Direct-reading ferrography
Emission spectroscopy
Infrared spectroscopy
Visual and odor
Percent solids/water
Viscosity measurement
Analytical ferrography
Total acid number (TAN)
Total base number (TBN)

One limitation of thermography is that it is limited to line of sight, and errors can be introduced due to material geometry, color of material, and environmental factors such as wind effects, solar heating, etc.

4. *Lubricant and wear particle analysis:* Three reasons for performing this type of analysis are: to assess wear condition of equipment, to assess the lubricant condition, and to assess if the lubricant is contaminated. The test used for the above purposes will depend on factors such as cost, sensitivity and accuracy of the test results, and the equipment construction and application.

   A list of standard analytical tests is presented in Table 6.4.

5. *Passive (airborne) ultrasonics:* Airborne ultrasonic devices (AVD) function within the frequency spectrum of 20 to 100 kHz and heterodyne the high frequency signal to the audible level so that the operator is able to hear changes in noise associated with leaks, corona discharges, etc. Two typical examples are bearing ring and housing resonant frequency excitation due to inadequate lubrication and minor defects.

   Some specific equipment application examples are: heat exchangers, boilers, and bearings. One of the main limitations of the airborne ultrasonics (AUs) technique is that AUs are subjective and dependent on perceived differences in noises.

6. *Nondestructive testing:* This technique can determine material properties and quality of manufacture for high-value parts/assemblies without damaging the product or its function. Usually, nondestructive testing (NDT) is practiced when approaches such as destructive testing are cost-prohibitive or ineffective. NDT is associated with welding of large high-stress parts such as pressure vessels and structural supports. In addition, oil refineries

and chemical plants use NDT methods to assure pressure boundaries' integrity for systems processing of volatile substances.

NDT techniques include: ultrasonic testing (imaging), magnetic particle testing, dye penetrant inspections, hydrostatic testing, eddy current testing, and radiography. Prior to implementation of an NDT program, it is recommended that a formal plan be developed. It would incorporate factors such as the technique to be used, number and orientation of samples, frequency, location, the failure mode each sample should address, and the information to be gained from each sample. The interval between inspections and the location of sampling points are two of the more difficult variables to address.

In the case of time interval between inspections, in establishing sample intervals or frequency, the factors that must be examined include system operating cycle, type of contained substance, major corrosion mechanisms, historical failure rate, expected corrosion rate, proximity of existing material to minimum wall thickness, erosion mechanisms, and expected erosion rate. Similarly in the case of location of sampling points, some of the guidelines for locating NDT sampling points are as follows:

- Welds, high stress fasteners, and stressed areas
- Areas susceptible to cavitation
- Dissimilar metals' junctions
- Areas with identified accelerated corrosion/erosion mechanisms
- Abrupt changes in direction of flow (elbows) and changes in pipe diameter
- "Dead-heads"

Table 6.5 presents application areas for specific NDT techniques. Limitations associated with each NDT technique are given in Table 6.6.

TABLE 6.5
Application Areas for Specific NDT Techniques

NDT Technique	Application Areas
Ultrasonic testing (imaging)	Metal components including weld deposits and specialized applications for plastics or composite materials
Dye penetrant inspections	Nonporous materials (those chemically compatible with the dye and developer)
Hydrostatic testing	Components and completely assembled systems containing pressurized fluids or gases
Radiography	Metal components including weld deposits and, possibly, specialized applications for plastics or composite materials
Magnetic particle testing	Materials that conduct electric current and magnetic lines of flux
Eddy current testing	Detect defects such as seams, cracks, holes, or lamination separation on both flat sheets and more complex cross-sections. Also, monitor the thickness of metallic sheets, plates, and tube walls

TABLE 6.6
NDT Technique Limitations

NDT Technique	Limitations
Ultrasonic testing (imaging)	One-dimensional, thus defects that parallel the axis of the test will not be apparent.
Dye penetrant inspections	Minute surface discontinuities such as machining marks will become readily apparent.
Hydrostatic testing	Over-pressurization can result in unintended damage to the system; cleanliness and fluid chemistry control must be compatible with system operating standards.
Radiography	Effective application requires expensive equipment, properly trained technicians to interpret images, and extensive safety precautions.
Magnetic particle testing	Applicable only to those materials that conduct electric current and influence magnetic lines of flux. Also, small areas between the two electrodes can only be inspected effectively.
Eddy current testing	Limited to shallow subsurface and surface defects. Also, due to the tendency of eddy currents to flow parallel to the surface to which the exciting field is applied, some laminar discontinuities' orientations parallel to this surface tend to remain undetected.

RCM PROGRAM EFFECTIVENESS MEASUREMENT INDICATORS

Over the years many management indicators to measure the effectiveness of an RCM program have been developed.[11] The numerical indicators or metrics are considered the most useful because they are objective, precise, quantitative, and more easily trended than words, as well as consisting of a descriptor and a benchmark. A descriptor may be defined as a word or group of words detailing the units, the function, and the process under consideration for measurement. A benchmark is a numerical expression of a set goal. Some of the metrics for measuring the effectiveness of an RCM program are presented below along with suggested benchmarks. These benchmarks are the averages of data surveyed from around 50 major corporations in the early 1990s.[11]

EQUIPMENT AVAILABILITY

This is expressed by

$$EA = \frac{H_{ea}}{TH_{rp}} \qquad (6.1)$$

where
   EA   = equipment availability,
   H_{ea} = number of hours each unit of equipment is available to run at capacity,
   TH_{rp} = total number of hours during the reporting period.

The benchmark figure for this metric is 96%.

EMERGENCY PERCENTAGE INDEX

This is defined by

$$EP = \frac{H_{ej}}{TH_w} \tag{6.2}$$

where
   EP   = emergency percentage,
   H_{ej} = total number of hours worked on emergency jobs,
   TH_w = total number of hours works.

The benchmark figure for this indicator is 10% or less.

PTI COVERED EQUIPMENT INDEX

This index is used to calculate the percent of candidate equipment covered by PTI and is expressed by

$$P_{epti} = \frac{E_i}{TE_c} \tag{6.3}$$

where
   P_{epti} = percent of candidate equipment covered by PTI,
   E_i = total number of equipment items in PTI program,
   TE_c = total number of equipment candidates for PTI.

The benchmark figure for this metric is 100%.

FAULTS FOUND IN THERMOGRAPHIC SURVEY INDEX

This is expressed by

$$P_{fft} = \frac{TFN}{DS} \tag{6.4}$$

where
    P_{fft} = percent of faults found in thermographic survey,
    TNF = total number of faults found,
    DS   = total number of devices surveyed.

The benchmark figure for this index is 3% or less.

MAINTENANCE OVERTIME PERCENTAGE INDEX

This is expressed by

$$P_{mo} = \frac{TMOH}{TRMH} \tag{6.5}$$

where
    P_{mo} = maintenance overtime percentage,
    TMOH = total number of maintenance overtime hours during period,
    TRMH = total number of regular maintenance hours during period.

The benchmark figure for this metric is 5% or less.

FAULTS FOUND IN STEAM TRAP SURVEY INDEX

This index is expressed by

$$P_{ffs} = \frac{DST}{STS} \tag{6.6}$$

where
    P_{ffs} = percent of faults found in steam trap survey,
    DST = total number of defective steam traps found,
    STS = total number of steam traps surveyed.

The benchmark figure for this index is 10% or less.

PM/PTI-REACTIVE MAINTENANCE INDEX

This index is divided into two areas: PM/PTI and reactive maintenance. The PM/PTI-related index is expressed by

$$P_{pp} = \frac{MHPP}{MHR + MHPP} \tag{6.7}$$

where
    P_{pp} = PM/PTI work percentage,
    MHPP = total manhours of PM/PTI work,
    MHR   = total manhours of reactive maintenance work.

The benchmark figure for this metric is 70%.

The reactive maintenance related index is defined by

$$P_{rm} = \frac{MHR}{MHR + MHPP} \tag{6.8}$$

where
P_{rm} = reactive maintenance work percentage.

The benchmark figure for this index is 30%. The sum of Eqs. (6.7) and (6.8) is equal to unity or 100%.

EMERGENCY-PM/PTI WORK INDEX

This is expressed by

$$P_{epp} = \frac{TEH}{TPPMH} \tag{6.9}$$

where
P_{epp} = percent of emergency work to PTI and PM work,
TEH   = total number of emergency work hours,
TPPMH = total number of PTI and PM work hours.

The benchmark figure for this metric is 20% or less.

RCM ADVANTAGES AND REASONS FOR ITS FAILURES

The application of RCM has many benefits, including improvement in safety and environmental protection, improvement in product quality, improvement in the useful life of costly items, a maintenance database, improvement in teamwork, improvement in maintenance cost-effectiveness, greater motivation of individuals, and higher plant availability and reliability.[8,15]

Occasionally, application of RCM has resulted in failure. Some reasons for its failure were: an analysis conducted at too low a level, too much emphasis placed on failure data, the application was superfluous or hurried, computers were used to drive the process, only one individual was assigned to apply RCM, only the maintenance department on its own applied RCM, and manufacturers/ equipment vendors were asked to apply RCM on their own.[8]

PROBLEMS

1. What are the principal goals of RCM?
2. Discuss at least ten basic principles of RCM.
3. Describe the RCM process.
4. What are the four major components of RCM?

5. Discuss in detail, the following items:
   • Reactive maintenance
   • Proactive maintenance
6. List eight basic techniques employed by proactive maintenance to extend equipment life.
7. Describe the following techniques used by proactive maintenance to extend equipment life:
   • Age exploration
   • Root-cause analysis
   • Recurrence control
8. Describe the following PTI technologies/approaches:
   • Electrical condition monitoring
   • Thermography
   • Vibration monitoring and analysis
9. What is nondestructive testing? Discuss at least five techniques associated with nondestructive testing.
10. Define the following indexes associated with RCM:
    • Emergency percentage index
    • Maintenance overtime percentage index
    • Equipment availability
11. What are the advantages of RCM?
12. List reasons for RCM application failure.

REFERENCES

1. McKenna, T. and Oliverson, R., *Glossary of Reliability and Maintenance Terms,* Gulf Publishing Co., Houston, Texas, 1997.
2. Picknell, J. and Steel, K.A., Using a CMMS to support RCM, *Maintenance Technology,* October 1997, 110–117.
3. AMC Pamphlet No. 750-2, *Guide to Reliability Centered Maintenance,* Department of the Army, Washington, D.C., 1985.
4. MSG1, *Maintenance Evaluation and Program Development, 747 Maintenance Steering Group Handbook,* Air Transport Association, Washington, D.C., 1968.
5. MSG2, *Airline/Manufacturer Maintenance Program Planning Document,* Air Transport Association, Washington, D.C., 1970.
6. MSG3, *Airline/Manufacturer Maintenance Program Planning Document,* Air Transport Association, Washington, D.C., 1980.
7. Anderson, R.T. and Neri, L., *Reliability Centered Maintenance: Management and Engineering Methods,* Elsevier Applied Science Publishers, London, 1990.
8. Moubray, J., *Reliability Centered Maintenance,* Industrial Press, New York, 1997.
9. August, J., *Applied Reliability Centered Maintenance,* Pennklell, Tulsa, Oklahoma, 1999.
10. Nowlan, F.S. and Heap, H.F., *Reliability Centered Maintenance,* Dolby Access Press, San Francisco, CA, 1978.
11. *Reliability Centered Maintenance Guide for Facilities and Collateral Equipment,* National Aeronautics and Space Administration (NASA), Washington, D.C., 1996.

12. Dhillon, B.S., *Engineering Maintainability,* Gulf Publishing Co., Houston, Texas, 1999.

13. Smith, A.M., *Reliability Centered Maintenance,* McGraw-Hill, New York, 1993.

14. Netherton, D., RCM tasks, *Maintenance Technology,* July/August 1999, 61–69.

15. Brauer, D.C. and Brauer, G.D., Reliability-centered maintenance, *IEEE Transac. Reliability,* 36, 1987, 17–24.

16. Dhillon, B.S. and Viswanath, H.C., Bibliography of literature on failure data, *Micro-electronics and Reliability,* 30, 1990, 723–750.

17. NAVAIR 00-25-403, *Guidelines for the Naval Aviation Reliability-Centered Maintenance Process,* Naval Air Systems Command, Department of Defense, Washington, D.C., October 1996.

7 Inventory Control in Maintenance

INTRODUCTION

Usually, the major complaint of those involved in maintenance is the unavailability of materials and spares at the moment of need. Today, as modern engineering equipment grows more complex, the cost of inventorying spares has increased alarmingly.

In many maintenance organizations, materials account for one-third to one-half of the operating budget, and more in some capital-intensive industrial sectors.[1] Needless to say, maintenance functions rely heavily on the availability of items such as spares for production equipment and machinery. Specific examples include lubricants, valves, pipe fittings, paints, angle iron, channel iron, controls, and nuts and bolts. A well-managed inventory system of such items helps reduce maintenance cost, worker and equipment downtime, and improves productivity. Inventory control plays an important role in maintenance.

The history of inventory control begins in the early 1900s, and it was used only for clerical help for first-line management.[2] Today, inventory control has risen to higher organizational levels with responsibilities in areas such as planning and control. This chapter discusses important aspects of inventory control related to maintenance.

INVENTORY PURPOSES, TYPES, AND BASIC MAINTENANCE INVENTORY-RELATED DECISIONS

Inventory can help organizations in many ways. Figure 7.1 presents six important purposes.[3-5] There are many types of inventory. The commonly identified types include raw materials inventory, finished goods inventory, supplies inventory, work-in-process (WIP) inventory, transportation inventory, and replacement parts inventory.[4]

In the case of raw materials inventory, items are purchased from suppliers for use in production processes. Finished goods inventory is concerned with finished product items not yet delivered to customers. The supplies inventory is concerned with parts/materials used to support the production process. Usually these items are not an element of the product.

The WIP inventory is concerned with partly-finished items (i.e., components, parts, subassemblies, etc.) that have been started in the production process but must be processed further. The transportation inventory is concerned with items being shipped from suppliers or to customers through the distribution channel. The replacement parts inventory is concerned with maintaining items for the replacement of other items in the company or its customer equipment/systems as they wear out.

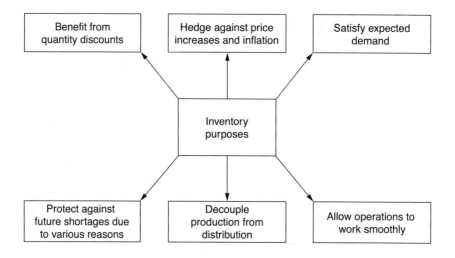

FIGURE 7.1 Important purposes of inventory.

Maintenance management personnel make decisions on basic areas such as those listed below with respect to inventory.[1]

- *Items/materials to be stored:* Decisions require consideration of factors such as ability of the vendor to supply at the moment of need, cost, and the degree of deterioration in storage.
- *Amount of items/materials to be stored:* Decisions are made by considering factors such as degree of usage and delivery lead time.
- *Item/material suppliers:* Decisions on suppliers of items/materials are made by considering factors such as price, delivery, quality, and service.
- *Lowest supply levels:* Decisions on lowest levels of supplies, in particular the major store items, are made by considering factors such as purchasing's historical records and projected needs.
- *Highest supply levels:* As time-to-time supply usage rate drops, the decisions on the highest supply levels are made by keeping in mind factors such as past ordering experience and peak vacation period.
- *Time to buy and pay:* Decisions on these two items are often interlocked. Such decisions are made by considering factors such as vendor announcements about special discounts, past purchasing records, and store withdrawals and equipment repair histories.
- *Place to keep items/materials:* As location control is crucial to a productive maintenance department, decisions concerning storage of items/materials are made by keeping in mind that they can be effectively retrieved. Past experience indicates that a single physical location for each item is the best.
- *Appropriate price to pay:* Pricing is of continuous concern, and decisions concerning it are primarily governed by perceived, not actual, supply and demand.

ABC CLASSIFICATION APPROACH FOR
MAINTENANCE INVENTORY CONTROL

In any maintenance inventory control system, parts/materials required for routine maintenance should be readily available. In the case of nonroutine maintenance, items must be controlled in such a way that the capital inventory investment is most effective.[6] In controlling inventory, one must seek information on areas such as those listed below.[7]

- Importance of the inventory item
- The way it should be controlled
- Quantity to be ordered at one time
- Specific point in time to place an order

The ABC classification approach provides information for routine and nonroutine maintenance. Consequently, it allows different levels of control based on the item's relative importance. The ABC approach is based on the reasoning that a small percentage of items usually dictates the results achieved under any condition. This reasoning is often referred to as Pareto principle, named after Vilfredo Pareto (1848–1923), an Italian sociologist and economist.[8]

The ABC approach classifies in-house inventory into three categories (i.e., A, B, and C) based on annual dollar volume. The following approximate relationship between the percentage of inventory items and the percentage of annual dollar usage is observed:[7,9]

- A: Of the items, 20% are responsible for 80% of the dollar usage.
- B: Of the items, 30% are responsible for 15% of the dollar usage.
- C: Of the items, 50% are responsible for 5% of the dollar usage.

The following three steps are associated with the ABC classification approach:

1. Determine the item characteristics that can influence inventory management results. Often, this is the annual dollar usage.
2. Group items based on the criteria established above.
3. Practice control relative to the group importance.

Factors such as annual dollar usage, material scarcity, and unit cost affect the importance of an item. Figure 7.2 presents the approach for grouping by annual dollar usage.[7]

CONTROL POLICIES FOR A, B, AND C CLASSIFICATION ITEMS

After the classification of inventory items, control policies can be established. Some of these policies associated with each classification are as follows:

- *Classification A items:* These are high-priority items. Practice tight control including: frequent review of demand forecasts, complete accurate records, periodic and frequent review by management, close followup, and expediting to minimize lead time.

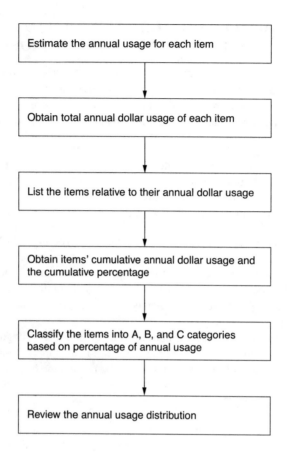

FIGURE 7.2 Steps for grouping by annual dollar usage.

- *Classification B items:* These are medium-priority items. Practice regular controls including: good records, regular processing, and normal attention.
- *Classification C items:* These are low-priority items. Practice simple controls, but ensure they are sufficient to meet demand.

Keep plenty of low-cost items, and use the money and control effort saved to minimize inventory of high-cost items.[7]

Example 7.1

A maintenance department uses ten types of items. Table 7.1 presents their annual usage and cost per unit. Determine the following:

- The annual dollar usage for each item
- The ordered list of items with respect to their annual dollar usage
- The cumulative yearly dollar usage and the cumulative percent of items
- The A, B, and C classifications of items

TABLE 7.1
Data for Ten Different Items

Item No.	Annual Usage (Units)	Cost per Unit ($)
1	400	10
2	200	40
3	1000	5
4	100	15
5	50	80
6	1700	5
7	500	10
8	600	50
9	700	200
10	900	4

TABLE 7.2
Annual Dollar Usage for Each Item

Item No.	Annual Usage (Units)	Annual Dollar Usage ($)
1	400	4,000
2	200	8,000
3	1000	5,000
4	100	1,500
5	50	4,000
6	1200	24,000
7	500	5,000
8	600	30,000
9	800	48,000
10	900	3,600
Total	5750	133,100

Table 7.2 presents the annual dollar usage for each item. Table 7.3 presents the ordered list of items with respect to their annual dollar usage, the cumulative annual dollar usage and the cumulative percent of items, cumulative percentage of dollar usage, and the A, B, and C classifications of items.

INVENTORY CONTROL MODELS

Various mathematical inventory control models have been developed, many of which can be applied to maintenance inventory control. These models are based on the assumption that demand for an individual item can be either independent of or dependent on the demand for other items. This section presents three models for managing independent demand items.

TABLE 7.3
A, B, and C Classifications of Ten Items

Item No.	Ordered Annual Dollar Usage	Cumulative Annual Dollar Usage	Cumulative Percentage of Dollar Usage	Cumulative Percentage of Items	Item Classification
9	140,000	140,000	66.79	10	A
8	30,000	170,000	81.10	20	A
6	8,500	178,500	85.16	30	B
2	8,000	186,500	88.98	40	B
3	5,000	191,500	91.36	50	B
7	5,000	196,500	93.75	60	C
5	4,000	200,500	95.66	70	C
1	4,000	204,500	97.57	80	C
10	3,600	208,100	99.28	90	C
4	1,500	209,600	100.00	100	C

TABLE 7.4
Inventory Holding Cost Elements[5]

Element	Sub-Elements	Approximate Cost Range as Percentage of Inventory Value (%)
Housing cost	Operating cost, building rent, insurance, depreciation, etc.	3–10
Investment cost	Cost of borrowing, cost of insurance on inventory, taxes, etc.	6–24
Labor cost (associated with extra handling)	—	3–5
Material handling cost	Equipment lease, power, operating costs, etc.	1–3.5
Miscellaneous cost	Costs associated with scrap, pilferage, obsolescence, etc.	2–5

Before we describe these models, let us first examine the following types of costs associated with the models:[5]

- *Holding cost:* This is associated with holding or carrying inventory over time. It also includes elements such as cost of insurance, extra staffing, and interest payments. In determining the holding cost, one must evaluate the kinds of costs shown in Table 7.4. It is emphasized that the figures given in the table are approximate and can vary substantially due to factors such as the nature of the business, current interest rate, and location.

- *Ordering cost:* This is associated with order processing, clerical support, forms, supplies, etc.
- *Setup cost:* This is associated with the preparation of an equipment/machine or process for manufacturing an order.

Note that often annual inventory cost approaches around 40% of the value of inventory.[5]

ECONOMIC ORDER QUANTITY MODEL

The economic order quantity model may be traced back to 1915 and is one of the most widely known inventory control methods.[10] Some assumptions associated with the model are as follows:

- Constant and known demand
- Instantaneous receipt of inventory
- Constant and known time between order placement and receipt of the order
- Infeasible quantity discounts
- Stockouts can be avoided by placing orders at the right time
- Two variable costs: holding cost and ordering or setup cost

The annual setup cost (SUC) is given by

$$SUC = NOP \times SOC = \frac{\theta}{q} \times SOC \qquad (7.1)$$

where
NOP = number of orders placed per year,
SOC = setup or ordering cost per order,
θ = demand in units for the inventory item annually,
q = number of pieces per order.

The annual inventory holding cost (AHC) is expressed by

$$AHC = AIL \times HC = \frac{q}{2} \times HC \qquad (7.2)$$

where
AIL = average inventory level,
HC = holding or carrying cost per unit per year.

For all optimal order quantity we equate Eqs. (7.1) and (7.2) as follows:

$$SUC = AHC = \frac{\theta}{q} \times SOC = \frac{q}{2} \times HC \qquad (7.3)$$

Solving Eq. (7.3), we get

$$q^* = \sqrt{\frac{2\theta \times \text{SOC}}{\text{HC}}} \tag{7.4}$$

where
q^* = optimum number of pieces per order or, specifically, the economic order quantity (EOQ).

The expected number of orders per year is given by

$$n = \frac{\theta}{q^*} \tag{7.5}$$

where
n = annual expected number of orders.

The expected time between orders (ETBO) is expressed by

$$\text{ETBO} = \frac{\text{TWD}}{n} \tag{7.6}$$

where
TWD = total number of working days in a year.

The daily demand (DD) is given by

$$\text{DD} = \frac{\theta}{\text{TWD}} \tag{7.7}$$

The reorder point (ROP) is expressed by

$$\text{ROP} = \text{DD} \times \text{LT} \tag{7.8}$$

where
LT = lead time for a new order expressed in days. Equation (7.8) is valid only if the demand is uniform and constant.

Example 7.2

A maintenance engineering department annually uses 600 units of a certain engineering part and the yearly holding cost per unit is $1.20 along with the setup or ordering cost of $5 per order. Calculate the following:

• Optimal number of units per order
• Expected number of orders per year
• Expected time between orders

By substituting the given data into Eq. (7.4), we get

$$q^* = \sqrt{\frac{2 \times 600 \times 5}{1.20}} \simeq 71 \text{ units}$$

The above value and the given data in Eq. (7.5) yields $n = 600/71 \simeq 8$ orders per year. Assuming 250 working days per year and the above calculated value, from Eq. (7.6) we get ETBO $= 250/8 \simeq 31$ days between orders. It means, there will be approximately 71 units per order, 8 orders per year, and 31 working days between each order.

Example 7.3

A maintenance organization uses 1000 units of a specific motor part annually and, on average, the delivery of an order takes 5 working days. If the organization operates a 250-day working year, calculate the reorder point by using Eq. (7.8).

By inserting the specified data into Eq. (7.7) we get DD $= 1000/250 = 4$ units per day. The above value and the given data in Eq. (7.8) yields ROP $= 4 \times 5 = 20$ units. It means that when the inventory level drops to 20 units an order should be executed.

PRODUCTION ORDER QUANTITY MODEL

In the economic order quantity model, it was assumed that the complete inventory order was received by the maintenance department at one time. However, there are instances when the department may receive its order over a period of time. To handle this case a different model is needed. The main assumption is that the manufacturer units cannot instantaneously produce all the units ordered. Consequently, this finite replenishment rate can impact the calculation of EOQ significantly.

Under the finite replenishment rate scenario, the inventory does not jump to the order quantity level at the occurrence of a replenishment because some items are being removed from the inventory at the time of replenishment. Consequently, the maximum inventory will never reach the level of the quantity ordered. The production order quantity model considers the time for producing the quantity ordered. Thus, the replenishment period is[4]

$$RT = \frac{q}{r} \tag{7.9}$$

where
RT $=$ replenish time or period,
r $=$ replenish rate expressed in units per day.

The usage, U, during the replenishment period is expressed by

$$U = \frac{q}{r} \times DD \tag{7.10}$$

The maximum inventory level, MIL, is given by

$$\text{MIL} = q - U = q - \frac{q}{r} \times \text{DD} = q\left(1 - \frac{\text{DD}}{r}\right) \qquad (7.11)$$

Consequently, the annual inventory holding cost is

$$\text{AHC} = \frac{\text{MIL}}{2} \times \text{HC} = \frac{q}{2}\left(1 - \frac{\text{DD}}{r}\right) \times \text{HC} \qquad (7.12)$$

By equating Eqs. (7.1) and (7.12) we get

$$\frac{\theta}{q} \times \text{SOC} = \frac{q}{2}\left(1 - \frac{\text{DD}}{r}\right) \times \text{HC} \qquad (7.13)$$

Solving Eq. (7.13) yields

$$q^* = \sqrt{\frac{2\theta \times \text{SOC}}{\left(1 - \frac{\text{DD}}{r}\right) \times \text{HC}}} \qquad (7.14)$$

Example 7.4

Assume that a manufacturer of a special system part used forecasted a sale of 1200 units for the next year. In addition, it was estimated that the daily demand for the part and its production will be 10 and 12 units, respectively. If the annual holding cost per unit and the setup cost per order are $1.50 and $6.00, respectively, calculate the optimal number of units for an order using Eq. (7.14).

By inserting the given data values into Eq. (7.14), we obtain

$$q^* = \sqrt{\frac{2 \times 1200 \times 6}{\left(1 - \frac{10}{12}\right) \times 1.50}} = 240 \text{ units}$$

Each order should be for 240 units of the special part.

QUANTITY DISCOUNT MODEL

Another assumption associated with the economic order quantity model is that the unit procurement price remains constant irrespective of the number of units purchased. In real life, this may or may not be true as many companies offer quantity discounts. Thus, the order quantity can influence the purchase price of a unit. As the discount quantity increases, the unit cost goes down but the holding cost goes up because of large orders.

In this case, the important trade-off is between the increased holding cost and the reduced unit cost. We write the total annual inventory cost as follows:

$$\text{TAIC} = \text{SUC} + \text{AHC} + \text{PC} = \frac{\theta}{q} \times \text{SOC} + \frac{q}{2} \times \text{HC} + \theta\, C_u \qquad (7.15)$$

where
TAIC = total annual inventory cost,
PC   = product cost,
C_u = unit cost expressed in dollars per units.

The optimum number of units per order is given by

$$q^* = \sqrt{\frac{2\theta\,(\text{SOC})}{i C_u}} \qquad (7.16)$$

where
iC_u = unit annual holding cost expressed as a percentage i of the unit price C_u.

Example 7.5

A company manufactures a certain engineering part and offers sale quantity discounts as per Table 7.5. The annual demand for the part is 4000 units along with the ordering cost of $40 per order. The inventory holding cost is 25% of the part or unit price. Determine the order quantity that will minimize the total inventory cost.

By substituting the given values into Eq. (7.16), we get the following three values for q^*:

$$q_1^* = \sqrt{\frac{2 \times 4000 \times 40}{0.25 \times 10}} \simeq 358 \text{ units per order}$$

$$q_2^* = \sqrt{\frac{2 \times 4000 \times 40}{0.25 \times 8}} \simeq 400 \text{ units per order}$$

$$q_3^* = \sqrt{\frac{2 \times 4000 \times 40}{0.25 \times 7}} \simeq 428 \text{ units per order}$$

TABLE 7.5
Sale Quantity Discount Schedule

No.	Units Ordered	Part or Unit Price ($)
1	0–499	10
2	500–999	8
3	1000 and over	7

TABLE 7.6
Total Costs per Order Quantities

No.	Part or Unit Price ($)	Order Quantity	Annual Cost of Total Units ($)	Annual Holding Cost ($)	Annual Ordering Cost ($)	Total Cost ($)
1	10	358	40,000	447	447	40,894
2	8	500	32,000	500	320	32,820
3	7	1000	28,000	875	160	29,035

where

$q_1{}^*$, $q_2{}^*$, and $q_3{}^*$ = optimal values of q for the part or unit price of $10, $8, and $7, respectively.

If we examine the values for $q_1{}^*$, $q_2{}^*$, and $q_3{}^*$, we note the values of $q_2{}^*$ and $q_3{}^*$ are below the allowable discount rates in Table 7.5. Consequently, they must be adjusted to 500 and 1000, respectively. By using the specified data in the above equations, we obtained the total cost for each of the order quantities as shown in Table 7.6. From Table 7.6, it is apparent that an order of 1000 units will minimize the total cost.

SAFETY STOCK

The main purpose of having the safety stock is to mitigate the risk of running out of items at the moment of need. One technique for providing safety stock is known as the "two-bin system." In this case, a fixed replenishment order is placed as soon as the stock level hits the preset reorder point. More specifically, the items are stored in two bins, the replenishment order is placed as soon as the first bin becomes empty, the items from the second bin are used until receiving the ordered items. The value of the preset reorder point depends upon factors such as the rate of demand and its associated variability, the stock out cost, and the lead time and its associated variability.[11] Reference 7 states the following five factors on which the safety stock required depends:

1. Reorder frequency
2. Desired level of service
3. Ability to forecast/control lead times
4. Demand variability during the lead time
5. Length of the lead time interval

As the uncertainty in demand raises the possibility of a stock out, the demand for items can be specified by means of a probability distribution. Past experience indicates that often demand during the lead time follows the normal probability distribution. Thus, the safety stock necessary to get a desired level of service is given by[4]

$$ST = z\sigma \qquad\qquad (7.17)$$

where
  ST = safety stock necessary to get a desired service level,
  σ = demand standard deviation during lead time,
  z = number of standard deviations from the mean value required to get desired
      level of service.

Consequently, the order point is expressed by

$$ODP = \mu + z\sigma \tag{7.18}$$

where
  ODP = order point,
  μ = mean demand during lead time.

The value of z is estimated according to the desired level of service.

Example 7.6

A maintenance department has normally distributed average demand for an item during the reordering lead time with mean and standard deviation of 50 and 4, respectively. Determine the size of the safety stock and reorder point at 95% service level.

  Using the given data, the normal distribution table, and Eq. (7.17), we get ST = $1.64 \times 4 \simeq 7$ items or units. Inserting the above calculated value and the given data into Eq. (7.18) yields ODP = $50 + 7 = 57$ items or units. The size of the safety stock must be 7 items or units and the order must be placed when there are 57 items per unit in inventory.

INCREASING OR DECREASING MAINTENANCE INVENTORY-ASSOCIATED FACTORS AND A MODEL FOR ESTIMATING SPARE PART QUANTITY

There are a number of factors that tend to increase the amount of maintenance-related inventory and, ultimately, the cost of maintenance. Careful consideration of these factors can help reduce inventory costs and, in turn, the cost of maintenance. Some of the factors are as follows:[12]

- Cost of production downtime
- Lack of parts standardization
- Poor attention to inventory or order quantities
- Maintenance scheduling requirements
- Existence of multiple storage depots
- Inadequate attention paid to the economics of quantity purchasing
- Undependable suppliers
- Nature and condition of facilities

The factors that tend to decrease maintenance-related inventory include good service from suppliers, infrequent equipment breakdown, availability of cash, and the cost associated with storeroom activity.

SPARE PART QUANTITY ESTIMATION MODEL

In maintenance activity, it is important to estimate the number of spare parts required for a system/equipment. This need directly influences the maintenance inventory.

Over the years various approaches have been used to determine the spare part quantity. Determination of spare part quantity depends on factors such as the reliability of the item under consideration, the number of items used in the system, and the probability of having a spare available when required.[13]

Often, the following equation, based on the Poisson distribution, is used to determine the spare part quantity:[14–16]

$$P_s = \sum_{j=0}^{n} \{[(-1)\ln e^{-q\lambda t}]^j e^{-q\lambda t}\}/j! \tag{7.19}$$

where
  P_s = probability of having a spare part available when needed,
  λ = part failure rate,
  t = time,
  n = number of spare parts carried in inventory,
  q = number of parts of a specific type used.

Time to time P_s is also referred to as the safety factor. It indicates the desired level of protection in estimating the need for spares. By examining Eq. (7.19), the higher the value of P_s, the greater the quantity of spares required and, in turn, the higher the purchasing and inventory maintenance costs.

Example 7.7

A piece of equipment has 30 parts of a specific type with a failure rate of 20 failures per million hours of operation. Assume that the equipment is operated continuously throughout the day and night and the spares are restocked every 4 months. Calculate the probability of having a spare part available when required, if only 4 spare parts are carried in inventory.

By substituting the given data into Eq. (7.19), we get

$$P_s = \sum_{j=0}^{n} \frac{\{[(-1)\ln e^{-(30)(20\times10^{-6})(4)(30)(24)}]^j e^{-(30)(20\times10^{-6})(4)(30)(24)}\}}{j!}$$

$$= \sum_{j=0}^{4} [\{1.728\}^j (0.1776)]/j!$$

$$= 0.1776 + 0.3069 + 0.2652 + 0.1527 + 0.0660$$

$$= 0.9684$$

There is an approximately 97% chance that a spare part will be available when required.

PROBLEMS

1. Write an essay on the need for having maintenance-related inventory.
2. What are the different types of inventory?
3. What is Pareto principle?
4. Describe the ABC classification approach.
5. What are the factors on which the requirement for safety stock depends?
6. List the factors that tend to increase the amount of maintenance-related inventory.
7. Assume that a system has 40 parts of a specific type with a failure rate of 3 failures per 1000 hours of operation. The system is operated continuously throughout the day and night, and the spares are restocked every 3 months. Calculate the probability of having a spare part available when required, if only 5 spare parts are carried in inventory.
8. Prove that in the case of an Economic Order Quantity Model, the optimum number of pieces per order is given by

$$q^* = \sqrt{\frac{2\theta \times \text{SOC}}{\text{HC}}} \tag{7.20}$$

   where
   θ = demand rate in units for the inventory item annually,
   SOC = setup or ordering cost per order,
   HC = holding or carrying cost per unit per order.
9. A maintenance department annually uses 400 units of a specific engineering part and the yearly holding cost per unit is $0.90 along with the setup or ordering cost of $4 per order. Determine the following:
   - Optimal number of units per order
   - Expected time between orders
   - Expected number of orders per year
10. Describe the following two types of models:
    - Production order quantity model
    - Quantity discount model

REFERENCES

1. Westerkamp, T.A., *Maintenance Manager's Standard Manual,* Prentice Hall, Paramus, New Jersey, 1997.
2. Greene, J.H., *Production and Inventory Control Handbook,* McGraw-Hill, New York, 1987.
3. Render, B. and Heizer, J., *Principles of Operations Management,* Prentice Hall, Upper Saddle River, New Jersey, 1997.
4. Vonderembse, M.A. and White, G.P., *Operations Management,* West Publishing Company, New York, 1996.
5. Heizer, J. and Render, B., *Production and Operations Management,* Prentice Hall, Upper Saddle River, New Jersey, 1996.

6. Niebel, B.W., *Engineering Maintenance Management,* Marcel Dekker, New York, 1994.
7. Arnold, J.R.T., *Introduction to Materials Management,* Prentice-Hall, Upper Saddle River, New Jersey, 1996.
8. Hayes, G.E. and Romig, H.G., *Modern Quality Control,* Bruce, Encino, California, 1977.
9. Dickie, H.F., ABC analysis, *Modern Manufacturing* (formerly *Factory Management and Maintenance*), July 1951, 20–25.
10. Harris, F.W., *Operations and Cost,* A.W. Shaw Company, Chicago, 1915.
11. Kelly, A. and Harris, M.J., *Management of Industrial Maintenance,* Newnes-Butterworths, London, 1979.
12. Martin, J.C., Maintenance stores and inventory control, in *Maintenance Engineering Handbook,* L.R. Higgins, ed., McGraw-Hill, New York, 1988, pp. 2–127, 2–144.
13. Blanchard, B.S., Verma, D., and Peterson, E.L., *Maintainability,* John Wiley & Sons, New York, 1995.
14. Ebel, G. and Lang, A., Reliability approach to the spare parts problem, in *Proceedings of the National Symposium on Reliability and Quality Control,* 1963, 130–134.
15. Von Alven, W.H., ed., *Reliability Engineering,* Prentice-Hall, Englewood Cliffs, New Jersey, 1964.
16. Blanchard, B.S., *Logistics Engineering and Management,* Prentice-Hall, Englewood Cliffs, New Jersey, 1981.

8 Human Error in Maintenance

INTRODUCTION

Humans play an important role during the equipment life cycle in the design, production, and operation and maintenance phases. Even though the degree of their role may vary from one type of equipment to another and from one equipment phase to another, human interaction is subject to deterioration because of human error. Human error may be defined as the failure to perform a specified task (or the performance of a forbidden action) that could lead to disruption of scheduled operations or result in damage to property and equipment.[1-2]

Some of the causes of human error include poor equipment design, poor work environment, poor work layout, improper work tools, inadequate training, and poorly written equipment maintenance and operating procedures.[3] Human error may be classified into six categories: design, assembly, inspection, installation, operation, and maintenance.[3,4] Maintenance error occurs due to incorrect repair or preventive actions. Two typical examples are incorrect calibration of equipment and application of the wrong grease at appropriate points of the equipment. It is usually an accepted fact that the occurrence of maintenance error increases as the equipment ages due to the increase in maintenance frequency.[4]

This chapter presents some important aspects of human error in maintenance.

FACTS AND FIGURES ON HUMAN ERROR IN MAINTENANCE

Some of the facts and figures associated with human error in maintenance are as follows:

- A study of electronic equipment revealed that 30% of failures were due to operation and maintenance error.[5] The breakdown of this statistic shows abnormal or accidental condition (12%), manhandling (10%), and faulty maintenance (8%).
- In 1993, a study of 122 maintenance occurrences involving human factors concluded that the categories of maintenance error breakdowns were incorrect installations (30%), omissions (56%), incorrect parts (8%), and other (6%).[6,7]

- A study of safety issues with regard to onboard fatality of worldwide jet fleet for the period 1982–1991 indicates that maintenance and inspection was the second most important safety issue with 1481 onboard fatalities.[7,8]
- It is estimated that each delayed aircraft costs an airline on average $10,000 per hour. A small reduction in the frequency of maintenance-induced delays caused by humans could be beneficial to airlines in terms of cost.[7]
- A study of maintenance operations among commercial airlines revealed that 40 to 50% of the time the elements removed for repair were not defective.[9]
- In 1979 in a DC-10 accident at O'Hare Airport in Chicago, 272 persons died because of improper maintenance procedures.[9]
- An incident involving the blowout-preventer (assembly of valves) at the Ekofisk oil field in the North Sea was the result of upside-down installation of the device. Cost of the incident was estimated to be around $50 million.[9]
- A study of maintenance errors in missile operations revealed various causes: loose nuts/fittings (14%), incorrect installation (28%), dials and controls (misread, misset) (38%), inaccessibility (3%), and miscellanewous (17%).[9,10]
- In 1983, an L-1011 aircraft departing Miami lost oil pressure in all three engines due to missing chip detector O-rings. A subsequent investigation traced the fault to poor inspection and supply procedure.[11]
- A study of various tasks such as adjust, align, and remove indicates a human reliability mean of 0.9871.[12] This means that one should expect errors by maintenance personnel on the order of 13 times in 1000 attempts.[9]

MAINTENANCE ERROR IN SYSTEM LIFE CYCLE AND BREAKDOWN OF MAINTENANCE PERSON'S TIME

The occurrence of maintenance error in the system life cycle, i.e., from the time of acceptance to the start of phase-out period is an important factor. As shown in Fig. 8.1, its contribution to the total human error that causes system failure is at least equal to that of operator error.[9,13]

Figure 8.1 presents approximate breakdowns of human error in a system life cycle. Note that as the system ages, maintenance errors increase dramatically.

A good understanding of time spent on various maintenance tasks can be useful to analyze maintenance errors. Over the years many studies have shown that the majority of time is spent on fault-diagnosis. According to one study,[9] the maintenance person's time for electronic equipment may be classified into three groups: diagnosis, remedial actions, and verification. Figure 8.2 shows breakdown percentages for these three groups.

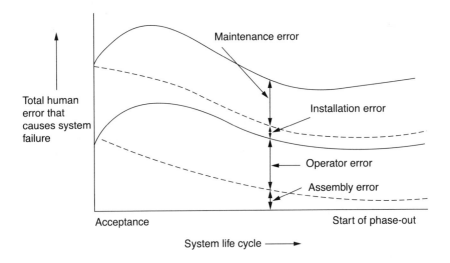

FIGURE 8.1 System life cycle vs. various types of human error.

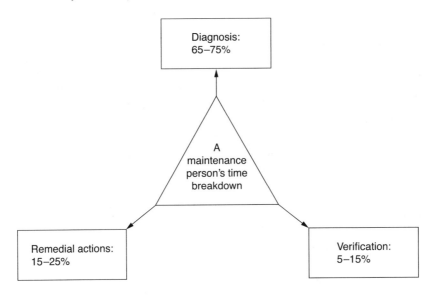

FIGURE 8.2 Breakdown percentages of a maintenance person's time.

TOP HUMAN FAILURE PROBLEMS IN MAINTENANCE, FREQUENCY OF MAINTENANCE ERROR TYPES, AND OUTCOMES OF MAINTENANCE INCIDENTS

There have been many studies of human factors in airline maintenance. According to one such study, the top eight human failures concerning maintenance in aircraft over 5700 kg were as shown in Fig. 8.3.[7] The human failures shown in Fig. 8.3 are

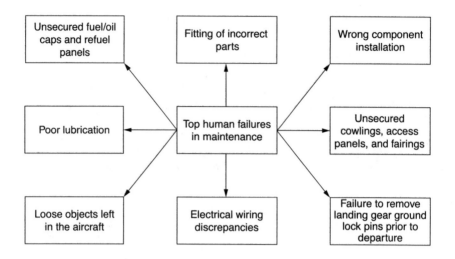

FIGURE 8.3 Top eight human failures in aircraft maintenance.

self-explanatory. The "electrical wiring discrepancies" include cross connections, and examples of "loose objects left in the aircraft" are various types of tools.

A study of 86 incident reports conducted by Boeing[14] classified maintenance errors into 31 categories along with their occurrence frequency. These maintenance errors with occurrence frequency shown in parentheses were: system operated in unsafe condition (16), towing event (10), system not made safe (10), equipment failure (10), degradation not found (6), falls and spontaneous actions (6), incomplete installation (5), work not documented (5), person entered dangerous area (5), person contacted hazard (4), system not reactivated/deactivated (4), did not obtain or use appropriate equipment (4), unserviceable equipment used (4), verbal warning not given (3), vehicle driving (not towing) (2), pin or tie left in place (2), warning sign or tag not used (2), not properly tested (2), safety lock or warning removed (2), vehicle/equipment contacted aircraft (2), material left in aircraft/engine (1), access panel not closed (1), contamination of open system (1), equipment not installed (1), panel installed incorrectly (1), required servicing not performed (1), unable to access part in stores (1), wrong equipment/part installed (1), wrong fluid type (1), wrong orientation (1), and miscellaneous (6). In the occurrence frequency, note that an incident may have involved more than one type of error.

In this study, the four most frequently occurring maintenance errors were system operated in unsafe condition, towing event, system not made safe, and equipment failure. Their combined frequency of occurrence was 46. System operated in unsafe condition error included incidents where aircraft systems such as thrust reversers or flaps were operated during maintenance actions in the presence of persons or obstructions in the area. The towing event error reflects the potential for damage to aircraft as they are maneuvred by maintenance persons within a restricted space.

The system not made safe error included situations where maintenance personnel failed to disable or lock out aircraft systems appropriately prior to commencement of maintenance work. For example, electrical power was left on during maintenance work.

The equipment failure error refers to situations where a maintenance equipment item or an aircraft component failed independent of maintenance actions. For example, unsafe behavior of maintenance persons resulted in their contact with faulty electrical equipment, thus nonfatal electric shocks.

A study of aircraft maintenance incidents revealed the following outcomes:[7]

- Potential hazard
- Correction of problem
- Exposure to hazard
- Potential damage to aircraft
- Damage to aircraft
- Aircraft signed off with unrectified problem
- Delayed aircraft
- Aircraft signed off with problem arising from maintenance action

The first four of the above outcomes are described below. The others are self-explanatory.

Potential hazard includes situations where there was a risk that workers could have been exposed to a hazard such as dangerous working surfaces or hydraulically activated aircraft components.

Correction of problem outcome refers to events where maintenance personnel committed errors, but prior to signing off from the work such errors were recognized and corrected.

Exposure to hazard includes situations where the resulting incident outcome was that a maintenance worker came into contact with a hazard, irrespective of his/her control over the event. One example of this is a maintenance worker cutting his/her hand when coming into contact with windmilling engine fan blades.

The potential damage to aircraft outcome refers to situations where workers failed to disable systems prior to carrying out maintenance work. Under such circumstance the systems would have been damaged if they had been turned on during maintenance.

REASONS FOR MAINTENANCE ERROR

There are many reasons for the occurrence of maintenance error. Some of those could be as follows:[9-10] poor work layout, poor equipment design, poorly written maintenance procedures, complex maintenance tasks, improper work tools, poor environment (i.e., temperature, humidity, lighting, etc.), fatigued maintenance personnel, outdated maintenance manuals, inadequate training and experience, etc.

With respect to training and experience, a study of maintenance technicians revealed that those who ranked highest possessed characteristics such as more experience, higher aptitude, greater emotional stability, fewer reports of fatigue, greater satisfaction with the work group, and higher morale.[9,12] Furthermore, correlation analysis indicated a significant degree of positive correlations between task performance and factors such as years of experience, amount of time in career field, responsibility-handling ability, and morale.

On the other hand, a significant degree of negative correlations were discovered between task performance and anxiety level and fatigue symptoms.

GUIDELINES FOR REDUCING HUMAN ERROR IN MAINTENANCE

Over the years guidelines have been developed to reduce human error in maintenance. This section presents guidelines developed to reduce human error in airline maintenance. Many of these guidelines can also be used in other areas of maintenance. The guidelines cover ten areas: procedures, human error risk management, tools and equipment, training, design, supervision, communication, shift handover, towing aircraft, and maintenance incident feedback.[7]

Procedures are covered by the following four guidelines:

1. Ensure, as much as possible, that standard work practices are followed all across maintenance operations.
2. Periodically review documented maintenance procedures and practices to ensure they are accessible, realistic, and consistent.
3. Periodically examine work practices to ensure that they do not differ significantly from formal procedures.
4. Evaluate the ability of checklists to assist maintenance persons in performing routine operations such as preparing an aircraft for towing, activating hydraulics, or moving flight surfaces.

There are three guidelines concerning human error risk management: (1) Carefully consider the need to disturb normally operating systems to perform nonessential periodic maintenance inspections, as there is risk of maintenance error occurrence associated with a disturbance, (2) Formally review the adequacy of defenses, such as engine runs, designed into the system for detecting maintenance errors, (3) Avoid, as much as possible, the simultaneous performance of the same maintenance task on similar redundant systems.

Two guidelines that cover tools and equipment are ensuring the storage of lockout devices in such a manner that it becomes immediately apparent if they are left in place inadvertently and reviewing the systems by which equipment, such as lighting systems and stands, is maintained for the removal of unserviceable equipment from service and repairing it rapidly.

The following guidelines are associated with training:

- Consider introducing crew resource management for maintenance professionals and others, i.e., persons interacting with the maintenance professionals.
- Offer periodic refresher training to maintenance professionals with emphasis on company procedures.

Two important guidelines concerning design are: ensure that manufacturers give proper attention to maintenance of human factors during the design process and actively seek information on the errors occurring during the maintenance phase for the input in the design phase.

A guideline in the area of supervision is to recognize that supervision and management oversight need to be strengthened, particularly in the final hours of each shift as the occurrence of errors becomes more likely.

In the area of communication, ensure that satisfactory systems are in place to disseminate important information to all maintenance staff so that changing procedures or repeated errors are considered in an effective manner. Shift handover can be a factor in maintenance error. One particular guideline concerns ensuring the adequacy of shift handover practices by carefully considering documentation and communication, so that incomplete tasks are transferred correctly across shifts.

In the area of towing aircraft or other equipment, review the procedures and equipment used for towing to and from maintenance facilities. Maintenance incident feedback is covered by the following guidelines:

- Ensure that management receives regular and structured feedback on maintenance incidents with particular consideration to the underlying conditions or latent failures that help promote such incidents.
- Ensure that engineering training schools receive regular feedback on recurring maintenance incidents so that effective corrective measures for these problems are targeted.

TECHNIQUES TO PREDICT THE OCCURRENCE OF HUMAN ERROR IN MAINTENANCE

Various techniques can be used to predict the probability of human error in maintenance. This section presents two such techniques: Markov and fault tree analysis (FTA) methods.

MARKOV METHOD

This method is used to perform reliability analysis of engineering systems and to predict the probability of occurrence of human error in maintenance. It is demonstrated through the following mathematical models. The method is described in detail in Chapter 12.

Model I

This mathematical model represents a system that can fail due to maintenance error or other failures.[15] Figure 8.4 shows the system transition diagram. Numerals in box, circle, and diamond denote system states. The following assumptions are associated with this model:

- The system can fail due to maintenance error or failures other than maintenance error.
- The failed system is repaired and preventive maintenance is performed periodically.

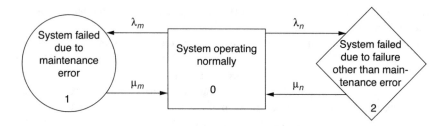

FIGURE 8.4 Maintenance error transition diagram.

- Maintenance error and failure, other than the maintenance error, rates are constant.
- Failed system repair rates are constant.
- The repaired system is as good as new.

The following symbols were used to develop equations for the model:

i = system state, $i = 0$ means the system operating normally, $i = 1$ means the system failed due to maintenance error, $i = 2$ means the system failed due to failure other than maintenance error,

t = time,

$P_i(t)$ = probability that the system is in state i at time t, for $i = 0, 1, 2,$

λ_m = system maintenance error rate,

λ_n = system nonmaintenance error failure rate,

μ_m = system repair rate from state 1,

μ_n = system repair rate from state 2.

Applying the Markov approach from Fig. 8.4 we obtain the following equations:

$$\frac{dP_0(t)}{dt} + (\lambda_m + \lambda_n)P_0(t) = \mu_m P_1(t) + \mu_n P_2(t) \qquad (8.1)$$

$$\frac{dP_1(t)}{dt} + \mu_m P_1(t) = \lambda_m P_0(t) \qquad (8.2)$$

$$\frac{dP_2(t)}{dt} + \mu_n P_2(t) = \lambda_n P_0(t) \qquad (8.3)$$

At $t = 0$, $P_0(0) = 1$, $P_1(0) = 0$, and $P_2(0) = 0$. By solving Eqs. (8.1)–(8.3), we get

$$P_0(t) = \frac{\mu_m \mu_n}{r_1 r_2} + \left[\frac{(r_1 + \mu_m)(r_1 + \mu_n)}{r_1(r_1 - r_2)}\right]e^{r_1 t} - \left[\frac{(r_2 + \mu_m)(r_2 + \mu_n)}{r_2(r_1 - r_2)}\right]e^{r_2 t} \qquad (8.4)$$

where

$$r_1, r_2 = \frac{-A \pm \sqrt{A^2 - 4(\mu_m \mu_n + \lambda_m \mu_n + \lambda_n \mu_n)}}{2}$$

$$A = \mu_m + \mu_n + \lambda_m + \lambda_n$$

$$r_1 r_2 = \mu_m \mu_n + \lambda_m \mu_n + \lambda_n \mu_m$$

$$r_1 + r_2 = -(\mu_m + \mu_n + \lambda_m + \lambda_n)$$

$$P_1(t) = \frac{\lambda_m \mu_n}{r_1 r_2} + \left[\frac{\lambda_m r_1 + \lambda_m \mu_n}{r_1(r_1 - r_2)}\right] e^{r_1 t} - \left[\frac{(\mu_n + r_2)\lambda_m}{r_2(r_1 - r_2)}\right] e^{r_2 t} \tag{8.5}$$

$$P_2(t) = \frac{\lambda_n \mu_m}{r_1 r_2} + \left[\frac{\lambda_n r_1 + \lambda_n \mu_m}{r_1(r_1 - r_2)}\right] e^{r_1 t} - \left[\frac{(\mu_m + r_2)\lambda_n}{r_2(r_1 - r_2)}\right] e^{r_2 t} \tag{8.6}$$

The probability of system failure due to maintenance error at time t is given by Eq. (8.5). The system availability is

$$AV_S(t) = P_0(t) = \frac{\mu_m \mu_n}{r_1 r_2} + \left[\frac{(r_1 + \mu_m)(r_1 + \mu_n)}{r_1(r_1 - r_2)}\right] e^{r_1 t} - \left[\frac{(r_2 + \mu_m)(r_2 + \mu_n)}{r_2(r_1 - r_2)}\right] e^{r_2 t} \tag{8.7}$$

where

AV_S (t) = system availability at time t.

As t becomes very large, the system steady-state availability from Eq. (8.7) is expressed by

$$AV_{SS} = \frac{\mu_m \mu_n}{r_1 r_2} \tag{8.8}$$

where

AV_{SS} = system steady-state availability.

Similarly, the steady-state probability of system failure due to maintenance error from Eq. (8.5) is

$$P_1 = \lim_{t \to \infty} P_1(t) = \frac{\lambda_m \mu_n}{r_1 r_2} \tag{8.9}$$

For $\mu_m = \mu_n = 0$ and regular preventive maintenance, from Eqs. (8.1)–(8.3) we get

$$P_0(t) = e^{-(\lambda_m + \lambda_n)t} \tag{8.10}$$

$$P_1(t) = \frac{\lambda_m}{\lambda_m + \lambda_n} [1 - e^{-(\lambda_m + \lambda_n)t}] \tag{8.11}$$

$$P_2(t) = \frac{\lambda_n}{\lambda_m + \lambda_n} [1 - e^{-(\lambda_m + \lambda_n)t}] \tag{8.12}$$

The probability of system failure due to maintenance error is given by Eq. (8.11). The system reliability is expressed by

$$R_S(t) = P_0(t) = e^{-(\lambda_m + \lambda_n)t} \tag{8.13}$$

where
$R_S(t)$ = system reliability at time t.

The system mean time to failure ($MTTF_S$) is given by

$$MTTF_S = \int_0^\infty R_S(t)\,dt$$

$$= \int_0^\infty e^{-(\lambda_m + \lambda_n)t}\,dt$$

$$= \frac{1}{\lambda_m + \lambda_n} \tag{8.14}$$

Example 8.1

Assume that a system can fail due to preventive maintenance-related human error and nonhuman error. The values of human and nonhuman error rates are 0.0002 errors per hour and 0.005 failures per hour, respectively. Calculate the probability of the system failure due to maintenance error for a 100-h mission.

By inserting the given values into Eq. (8.11) yields

$$P_1(100) = \frac{0.0002}{0.0002 + 0.005}\,[1 - e^{-(0.0002 + 0.005) \times 100}]$$

$$= 0.0156$$

There is approximately 1.6% chance the system will fail due to maintenance error.

Model II

This mathematical model represents a system that can only fail due to nonmaintenance-related failures, but its performance is degraded by the occurrence of maintenance error. The system transition diagram is shown in Fig. 8.5. Numerals in circle, box, and diamond denote system states.

The following assumptions pertain to this model:

- The totally or partially failed system is repaired and preventive mainte-nance is performed regularly.
- The occurrence of maintenance error can only lead to system degradation, but not failure.
- The system can fail from its degradation mode due to failures other than maintenance errors.

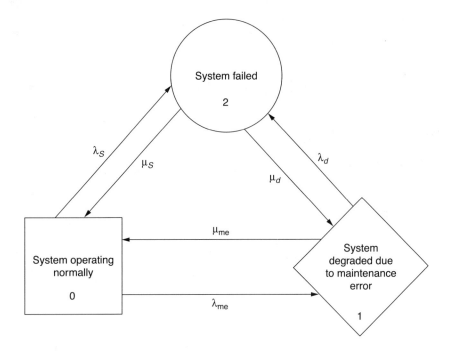

FIGURE 8.5 System state space diagram.

- The system is repaired at constant rates from its failed and degradation states.
- Maintenance error and nonmaintenance error failure rates are constant.
- The repaired system is as good as new.

The notations associated with this model are as follows:

i = system state, $i = 0$ means the system is operating normally, $i = 1$ means system is degraded due to maintenance error, $i = 2$ means the system failed,

$P_i(t)$ = probability that the system is in state i at time t, for $i = 0, 1, 2$,

λ_{me} = system maintenance error rate,

λ_S = system failure rate,

λ_d = system failure rate when in degradation mode,

μ_{me} = system repair rate from degradation mode,

μ_i = failed system ith repair rate, $i = S$ (to fully operational), $i = d$ (to degraded mode).

With the aid of the Markov method and Fig. 8.5, we write the following equations:

$$\frac{dP_0(t)}{dt} + (\lambda_{me} + \lambda_S)P_0(t) = \mu_{me}P_1(t) + \mu_S P_2(t) \qquad (8.15)$$

$$\frac{dP_1(t)}{dt} + (\mu_{me} + \lambda_d)P_1(t) = \mu_d P_2(t) + P_0(t)\lambda_{me} \tag{8.16}$$

$$\frac{dP_2(t)}{dt} + (\mu_S + \mu_d)P_2(t) = \lambda_d P_1(t) + \lambda_S P_0(t) \tag{8.17}$$

At $t = 0$, $P_0(0) = 1$, $P_1(0) = 0$, $P_2(0) = 0$.

By solving Eqs. (8.15)–(8.17), we obtain

$$P_0(t) = \frac{\mu_{me}\mu_S + \lambda_d\mu_S + \mu_{me}\mu_d}{K_1 K_2}$$

$$+ \left[\frac{\mu_{me}K_1 + \mu_S K_1 + \mu_d K_1 + K_1\lambda_d + K_1^2 + \mu_{me}\mu_S + \lambda_d\mu_S + \mu_{me}\mu_d}{K_1(K_1 - K_2)}\right]e^{K_1 t}$$

$$+ \left\{1 - \left(\frac{\mu_{me}\mu_S + \lambda_d\mu_S + \mu_{me}\mu_d}{K_1 K_2}\right)\right.$$

$$\left. - \left[\frac{\mu_{me}K_1 + \mu_S K_1 + \mu_d K_1 + K_1\lambda_d + K_1^2 + \mu_{me}\mu_S + \lambda_d\mu_S + \mu_{me}\mu_d}{K_1(K_1 - K_2)}\right]\right\}e^{K_2 t}$$

$$\tag{8.18}$$

where

$$K_1, K_2 = [-B \pm \sqrt{B^2 - 4(\mu_{me}\mu_S + \lambda_d\mu_S + \mu_{me}\mu_d + \mu_S\lambda_{me} + \lambda_{me}\mu_d + \mu_{me}\lambda_d + \mu_{me}\lambda_S + \lambda_S\mu_d + \lambda_S\lambda_d)}]/2$$

$$B = \lambda_{me} + \lambda_S + \lambda_d + \mu_{me} + \mu_S + \mu_d$$

$$K_1 K_2 = \mu_{me}\mu_S + \lambda_d\mu_S + \mu_{me}\mu_d + \mu_S\lambda_{me} + \lambda_{me}\mu_d + \lambda_{me}\lambda_d + \mu_{me}\lambda_S + \lambda_S\mu_d + \lambda_S\lambda_d$$

$$P_1(t) = \frac{\lambda_{me}\mu_S + \lambda_{me}\mu_d + \lambda_S\mu_d}{K_1 K_2} + \left[\frac{K_1\lambda_{me} + \lambda_{me}\mu_S + \lambda_{me}\mu_d + \lambda_S\mu_d}{K_1(K_1 - K_2)}\right]e^{K_1 t}$$

$$- \left[\frac{\lambda_{me}\mu_S + \lambda_{me}\mu_d + \lambda_S\mu_d}{K_1 K_2} + \frac{K_1\lambda_{me} + \lambda_{me}\mu_S + \lambda_{me}\mu_d + \lambda_S\mu_d}{K_1(K_1 - K_2)}\right]e^{K_2 t} \tag{8.19}$$

$$P_2(t) = \frac{\lambda_{me}\lambda_d + \mu_{me}\lambda_S + \lambda_S\lambda_d}{K_1 K_2} + \left[\frac{K_1\lambda_S + \lambda_{me}\lambda_d + \lambda_S\mu_{me} + \lambda_S\lambda_d}{K_1(K_1 - K_2)}\right]e^{K_1 t}$$

$$- \left[\frac{\lambda_{me}\lambda_d + \mu_{me}\lambda_S + \lambda_S\lambda_d}{K_1 K_2} + \frac{K_1\lambda_S + \lambda_{me}\lambda_d + \mu_{me}\lambda_S + \lambda_S\lambda_d}{K_1(K_1 - K_2)}\right]e^{K_2 t} \tag{8.20}$$

The probability of system degradation due to maintenance error is given by Eq. (8.19). As t becomes very large, it is expressed by

$$P_{1SS} = \frac{\lambda_{me}\mu_s + \lambda_{me}\mu_d + \lambda_s\mu_d}{K_1 K_2} \tag{8.21}$$

where
P_{1SS} = steady-state probability of system degradation due to maintenance error.

The time dependent system operational availability is given by

$$AV_{0S}(t) = P_0(t) + P_1(t) \tag{8.22}$$

where
$AV_{0S}(t)$ = system operational availability at time t.

As t becomes very large, the system operational availability from Eq. (8.22) is

$$AV_{0SS} = \frac{\mu_{me}\mu_s + \lambda_d\mu_s + \mu_{me}\mu_d + \lambda_{me}\mu_s + \lambda_{me}\mu_d + \lambda_s\mu_d}{K_1 K_2} \tag{8.23}$$

Example 8.2

Assume that for a system the following data are specified:

$\lambda_s = 0.004$ failures per hour, $\mu_s = 0.02$ repairs per hour,
$\lambda_{me} = 0.0001$ errors per hour, $\mu_{me} = 0.008$ repairs per hour,
$\lambda_d = 0.001$ failures per hour, $\mu_d = 0.06$ repairs per hour.

Calculate the steady-state probability of system degradation due to maintenance error by using Eq. (8.21).

Substituting the given values into Eq. (8.21), we get

$$
\begin{aligned}
P_{1SS} = &\frac{(0.0001 \times 0.02) + (0.0001 \times 0.06) + (0.004 \times 0.06)}{(0.008 \times 0.02) + (0.001 \times 0.02) + (0.008 \times 0.06) + (0.02 \times 0.0001)} \\
&+ (0.0001 \times 0.06) + (0.0001 \times 0.001) + (0.008 \times 0.004) \\
&+ (0.004 \times 0.06) + (0.004 \times 0.0001) \\
= &\, 0.2627
\end{aligned}
$$

The steady state probability of system degradation due to maintenance error is 0.2627.

FAULT TREE ANALYSIS METHOD

The fault tree analysis (FTA) method is used to perform reliability and safety analyses of engineering systems. It was originally developed to analyze the Minuteman Launch Control System in the early 1960s at Bell Laboratories. It can also be used to perform analysis of human error in maintenance. The following examples demonstrate the application of FTA in maintenance with respect to human error.

Example 8.3

Assume that an engineering system can fail due to a maintenance error caused by factors such as poor equipment design, inadequate training, poor work environment, use of deficient maintenance manuals, or inadequate work tools. Two major factors for poor equipment design are oversight or no formal consideration of the occurrence of maintenance error. The "no formal consideration to the occurrence of maintenance error" factor can be caused by either no requirement in design specifications or insufficient allocated funds.

Two important factors for poor work environment are inadequate lighting and inaccessibility. Similarly, two main causes for the use of deficient maintenance manuals are unavailability of compatible maintenance manuals or poorly written maintenance procedures.

Develop a fault tree for the top event "Engineering system failed due to maintenance error." Figure 8.6 shows a fault tree for the example.

Example 8.4

Assume that the probability of occurrence of events $E_1, E_2, ..., E_9$ shown in Fig. 8.6 is 0.02. For independent events, calculate the probability of occurrence of the top event T, i.e., engineering system failed due to maintenance error.

The probability of occurrence of event I_1 is

$$P(I_1) = P(E_1) + P(E_2) - P(E_1)P(E_2)$$
$$= 0.02 + 0.02 - (0.02 \times 0.02)$$
$$= 0.0396$$

The probability of occurrence of event I_2 is given by

$$P(I_2) = P(I_1) + P(E_7) - P(I_1)P(E_7)$$
$$= 0.0396 + 0.02 - (0.0396 \times 0.02)$$
$$= 0.0588$$

The probability of occurrence of event I_3 is

$$P(I_3) = P(E_6) + P(E_5) - P(E_6)P(E_5)$$
$$= 0.02 + 0.02 - (0.02 \times 0.02)$$
$$= 0.0396$$

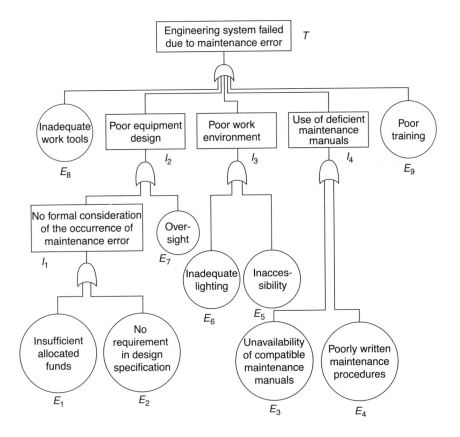

FIGURE 8.6 A fault tree for Example 8.3.

The occurrence probability of event I_4 is given by

$$P(I_4) = P(E_3) + P(E_4) - P(E_3)P(E_4)$$
$$= 0.02 + 0.02 - (0.02 \times 0.02)$$
$$= 0.0396$$

The top event T probability of occurrence is

$$P(T) = 1 - \{1 - P(E_8)\}\{1 - P(E_9)\}\{1 - P(I_2)\}\{1 - P(E_3)\}\{1 - P(I_4)\}$$
$$= 1 - (1 - 0.02)(1 - 0.02)(1 - 0.0588)(1 - 0.0396)(1 - 0.0396)$$
$$= 0.1663$$

There is approximately 17% chance the engineering system will fail due to maintenance error.

PROBLEMS

1. Write an essay on the occurrence of human error in engineering maintenance.
2. Discuss the occurrence of maintenance error during the system life cycle.
3. What are the typical breakdown percentages of a maintenance person's time?
4. List important human failures in aircraft maintenance.
5. What are the reasons for the occurrence of maintenance error?
6. List ten guidelines for reducing human error in maintenance.
7. Discuss techniques that can be used to predict the occurrence of human error in maintenance.
8. Assume that in Fig. 8.4, $\lambda_m = 0.0001$ errors per hour, $\lambda_n = 0.004$ failures per hour, $\mu_m = 0.04$ repairs per hour, and $\mu_n = 0.05$ repairs per hour. Calculate, the steady-state probability of system failure due to maintenance error by using Eq. (8.9).
9. Assume that in question 8, $\mu_m = \mu_n = 0$. Calculate the probability of system failure due to maintenance error for a 150-h mission by using Eq. (8.11).
10. Assume that in Fig. 8.5, $\mu_{me} = \mu_s = \mu_d = 0$. Develop an expression for the system mean time to failure.

REFERENCES

1. Meister, D., Human factors in reliability, in *Reliability Handbook*, W.G. Ireson, ed., Mcgraw-Hill, New York, 1966, 12.2–12.37.
2. Hagen, E.W., ed., Human reliability analysis, *Nuclear Safety,* 17, 1976, 315– 326.
3. Meister, D., The problem of human-initiated failures, in *Proceedings of the 8th National Symposium on Reliability and Quality Control,* 1962, 234–239.
4. Meister, D., *Human Factors: Theory and Practice,* John Wiley & Sons, New York, 1976.
5. AMCP 706-134, *Maintainability Guide for Design,* U.S. Army Material Command, Department of the Army, Washington, D.C., 1972.
6. Circular 243-AN/151, *Human Factors in Aircraft Maintenance and Inspection,* International Civil Aviation Organization, Montreal, Canada, 1995.
7. *Human Factors in Airline Maintenance: A Study of Incident Reports,* Bureau of Air Safety Investigation, Department of Transport and Regional Development, Canberra, Australia, 1997.
8. Russell, P.D., Management strategies for accident prevention, *Air Asia,* 6, 1994, 31–41.
9. Christensen, J.M. and Howard, J.M., Field experience in maintenance, in *Human Detection and Diagnosis of System Failures,* J. Rasmussen and W.B. Rouse, eds., Plenum Press, New York, 1981, 111–133.
10. Dhillon, B.S., *Human Reliability: With Human Factors,* Pergamon Press, New York, 1986.
11. Tripp, E.G., Human factors in maintenance, *B/CA,* July 1999, 1–10.
12. Sauer, D., Campbell, W.B., Potter, N.R., and Askren, W.B., Relationships between human resource factors and performance on nuclear missile handling tasks, Report

No. AFHRL-TR-76-85/AFWL-TR-76-301, Air Force Human Resources Laboratory/ Air Force Weapons Laboratory, Wright-Patterson Air Force Base, Ohio, 1976.

13. Rigby, L.V., The Sandia human error rate bank (SHERB), Report No. SC-R-67-1150, Sandia Labs, Albuquerque, New Mexico, 1967.

14. *Maintenance Error Decision Aid (MEDA),* Developed by Boeing Commercial Airplane Group, Seattle, Washington, 1994.

15. Dhillon, B.S., *Robot Reliability and Safety,* Springer-Verlag, New York, 1991.

9 Quality and Safety in Maintenance

INTRODUCTION

Quality may be defined as conformance to requirements or degree to which a product, function, or process satisfies the needs of customers and users.[1] Maintenance quality assurance is the actions by which it is determined that parts, equipment, or material maintained, modified, rebuilt, overhauled, or reclaimed conform to the specified requirements.[2] Maintenance quality is important because it provides a degree of confidence that maintained or repaired parts/equipment/systems will operate reliably and safely.[3]

In the United States, there is a fatal work-related injury every 103 minutes and a disabling injury every 8 seconds. In 1998 the total cost of work injuries was in the order of $125.1 billion.[4] Furthermore, unintentional injuries are the fifth leading cause of death, with an estimated cost of $480.5 billion per year.[4] Accidents occurring during maintenance work or concerning maintenance are frequent. For example, in 1994, 13.61% of all accidents in the U.S. mining industry occurred during maintenance work and, since 1990, the occurrence of such accidents has increased each year. It is essential that maintenance engineering should strive to eliminate or control potential safety hazards to ensure satisfactory protection to people and material from such things as electrical shock, high noise levels, fire and radiation sources, toxic gas sources, protruding structural members, and moving mechanical assemblies.[5]

This chapter presents important aspects of quality and safety in engineering maintenance.

NEED FOR QUALITY MAINTENANCE PROCESSES

Quality maintenance processes are an important factor in mission capability and personnel safety of many systems. Three examples presented below clearly demonstrate that even insignificant actions can lead to severe consequences. In all three cases, a strong and effective quality process may have averted tragedy.

- In 1990, a serious steam leak occurred in the fire room on the U.S.S. Iwo Jima (LPH2), a naval ship, that led to ten fatalities.[6] The investigation revealed that the cause was ship service turbine generator root valve bonnet fastener failure. A further investigation revealed that the valve had just been repaired, but the bonnet fasteners were replaced with mismatched and wrong material. The required fasteners were heat-treated steel studs and nuts, but the replacements were a mixture of bolts, studs, and black oxide-coated brass nuts. Ship's personnel furnished the replacement fasteners without proper verification that the requirements of the technical manual and drawings were satisfied.

- In 1986, the space shuttle Challenger exploded and seven astronauts lost their lives.[6,7] The cause of this disaster was identified as failure of the pressure seal in the aft field joint of the right solid rocket motor. Although the unexpected occurrences of O-ring erosion and blow-by were experienced often during shuttle flight history, neither the National Aeronautics Space Administration (NASA) nor the rocket engine builder developed a solution for the problem. It was concluded that a quality maintenance program would have tracked and found the cause for increasing erosion and blow-by.
- In 1963, a U.S. Navy nuclear submarine, the U.S.S. Thresher, was lost at sea off the coast of Maine because of a flooding casualty in its engine room.[6,8] An investigation concluded that the most likely cause was a piping failure in one of the salt water systems. Numerous changes were recommended in the design and maintenance processes for submarines.

MAINTENANCE WORK QUALITY

Good quality maintenance work leads to good results: reduction or elimination of unexpected failures, lower costs, better safety, increased confidence in work performed, etc. Good quality maintenance work can only be accurately measured after the specification of expectations. Once the aim of maintenance work is clearly identified, steps such as those listed below can be useful in producing good quality maintenance work.[9]

- *Limit perplexity.* Often the request for maintenance is incomplete and inaccurate. Ensure the work's proper completeness and accuracy prior to taking concrete steps.
- *Define goals.* Goals should be set by the maintenance team and its supervisors. Ensure that the team clearly understands the objectives for the maintenance work prior to its start.
- *Avoid unsafe practices.* Do not permit temptation to minimize maintenance time by short-cutting prescribed safety procedures.
- *Do not overlook secondary damage.* Ensure that less dramatic secondary problems are not overlooked. Otherwise, they could be costly at a later stage.
- *Report as the maintenance work progresses.* Report all relevant information that could be useful for performing similar tasks in future.
- *Do not use second-hand parts.* Ensure that failed parts are not replaced with second-hand parts.
- *Reinstall with extra care.* Due to various factors, the condition of some equipment/system parts or materials may deteriorate with time; thus when new or repaired parts/materials are reinstated, excessive force can damage other parts. Avoid introducing new failures while correcting the old ones.

- *Follow a system approach to box up.* There is tendency to close up quickly after finishing a repair. After the repair, it is important to consider the following factors:
  - Check for safety. Ensure that all hot-keys are returned to appropriate places and involved persons accounted for.
  - Check for all repair tools/equipment used. Do not restart in the event of missing items.
- *Test the repaired item prior to its hand-back.*
- *Complete all appropriate job records.* Tasks such as equipment, maintenance planning, and failure analysis rely heavily on an effective maintenance history.

QUALITY CONTROL CHARTS FOR USE IN MAINTENANCE

Walter A. Shewhart developed control charts in 1924.[10,11] Today such charts are used for many purposes: to determine whether or not the process is in the state of control, to provide information for decisions concerning inspection procedures or product specifications, etc. The control chart can be described as a graphical method used to evaluate whether a given process is in a "state of statistical control" or out of control. When a sampled value is outside the upper and lower control limits, it indicates that the process is not in the "state of statistical control," and thus warrants an immediate investigation to determine the cause for being out of control and appropriate subsequent corrective measures.

These control charts can be used in maintenance work to control the quality of the work performed. There are many different types of quality control charts: c-chart, p-chart, r-chart, etc.[12,13] The c-chart is described below.

C-CHART

In maintenance work, this chart can be used to control the occurrence of a number of maintenance-related equipment defects. Poisson distribution is used to obtain expressions for the upper and lower control limits for the chart. The Poisson distribution mean $\bar{C}$ is expressed by

$$\bar{C} = \frac{MD}{N} \tag{9.1}$$

where
  MD = total number of maintenance related defects,
  N = total number of equipment.

The standard deviation is

$$\sigma = \sqrt{\bar{C}} \tag{9.2}$$

TABLE 9.1
Maintenance-Related Defects
Associated with Each Machine

Machine No.	No. of Maintenance-Related Defects
1	2
2	6
3	1
4	5
5	3
6	0
7	6
8	4
9	1
10	5
11	1
12	3
13	2
14	4
15	2

The upper and lower control limits are

$$\text{UCL} = \bar{C} + 3\sigma \tag{9.3}$$

and

$$\text{LCL} = \bar{C} - 3\sigma \tag{9.4}$$

where
    UCL = upper control limit,
    LCL = lower control limit.

Example 9.1

A sample of 15 machines that underwent maintenance was examined for maintenance-related defects. Table 9.1 presents maintenance-related defects associated with each machine. Construct the c-chart using the specified data.

By summing the defects of Table 9.1, we get the following total: MD = 40. Inserting the above value and the other specified data into Eq. (9.1) yields $\bar{C}$ = 45/15 = 3. Thus, the average number of maintenance-related defects per machine is 3.

Using the above result in Eq. (9.2) yields $\sigma = \sqrt{3} = 1.7321$. Inserting the above calculated values in Eqs. (9.3) and (9.4), we get

$$\text{UCL} = 3 + 3(1.7321) = 8.1962$$

and

$$\text{LCL} = 3 - 3(1.7321) = -2.1962$$

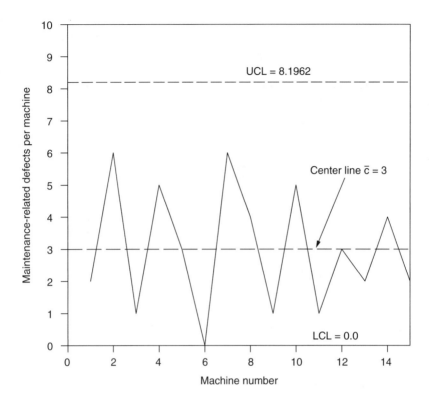

FIGURE 9.1 The c-chart.

As the calculated value of LCL = −2.1962 is impossible for plotting purposes, it is changed to zero, i.e., LCL = 0. Figure 9.1 shows a c-chart for the above-calculated values and the specified data. All the plotted data points fall within the upper and lower control limits. It means there is no abnormality in the occurrence of maintenance-related defects.

POST-MAINTENANCE TESTING*

Post-maintenance testing (PMT) helps increase the quality of maintenance performed. Basically, PMT has three objectives:[14]

1. Ensure that the original deficiency has been rectified appropriately.
2. Ensure that no new deficiencies have been introduced.
3. Ensure that the item under consideration is ready to perform its specified mission.

---

* Note that the material presented in this section is based on post-maintenance testing at nuclear facilities.[14] However, in its present form or with some modifications it can also be used for other purposes.

PMT should be carried out after all types of corrective maintenance activities as well as after some preventive maintenance actions as considered appropriate. Testing should be commensurate with the specific type of maintenance accomplished and the importance of structure/system/part to facility reliability and safety. In some situations, this may only require verification and checkout, but in others formal documented PMT may be necessary.

PMT Key Elements

PMT involves several key elements including:[14]

- Clearly defined responsibilities of each group involved
- Availability of guidance to planners to identify appropriate tests
- Scope of equipment tested incorporates all facility equipment/systems
- Performance of tests under the relevant system operational parameters
- Specification of appropriate tests incorporates inputs received from operating, maintenance, and technical support groups
- Testing performed with the consent of operator/owner, using authorized instructions/procedures, and conducted and reviewed by competent persons
- A form is used to document, authorize, and review PMT results

Operator-Documented PMT Responsibilities

The responsibilities of operator or owner of items/equipment requiring documented PMT include:[14]

- Defining the need for a level of PMT as well as for document approval
- Minimizing the redundancy of PMT and the PMT excessiveness
- Providing assistance to maintenance activity as required by performing applicable testing
- Clearly defining operational parameters and associated criteria
- Emphasizing the importance of ensuring configuration management
- Ensuring the proper authorization, performance, examination, and documentation of PMT before returning the equipment/item to operation
- Ensuring the performance of all associated delay tests before or in conjunction with returning the equipment/item to operation
- Restoring structures, systems, and components to exact set points for active or standby modes subsequent to testing

Types of Maintenance Activities/Items for PMT

There are various types of maintenance activities in which PMT could be of value. Some examples of those activities are listed below:[14]

- Maintenance that affects the operation or integrity of a gas/fluid system or parts within such systems

- Test and measurement equipment/system
- Maintenance that affects electrical distribution equipment, e.g., high-voltage or breaker connections
- Chemistry- and health physics-associated instrumentation
- Maintenance that affects the strength of parts/components/fittings
- Maintenance that affects electronic components or control circuitry, e.g., limit switches, controllers, protection relays
- Maintenance that affects parts in an instrument loop or instrument detectors

COMMON PMT ACTIVITIES

Some representative PMT activities are as follows:[14]

- Leak rate testing
- Current, voltage, or integrity checks
- Component calibration or alignment
- Nondestructive tests as specified by code
- Visual or dimensional inspections
- Part operational exercise including pressure, temperature, flow, and vibration
- Response time

MAINTENANCE SAFETY-RELATED FACTS, FIGURES, AND EXAMPLES, AND REASONS FOR SAFETY PROBLEMS IN MAINTENANCE

Some maintenance safety-related facts and figures are as follows:

- In 1998, approximately 3.8 million workers in the United States suffered from disabling injuries on the job.[4]
- In 1998, the total cost of work injuries in the United States was in the order of $125.1 billion.[4]
- In 1993, there were approximately 10,000 work deaths in the United States.[5]
- In 1994, 13.6% of the accidents in the United States mining industry occurred during maintenance.
- In 1991, an explosion at an oil refining company in Louisiana that killed four workers occurred as three gasoline synthesizing units were being brought back to their active state after some maintenance actions.[16]
- In 1990, a steam leak occurred in the fire room on board the U.S.S. Iwo Jima (LPH 2) naval ship, resulting in ten fatalities. Subsequently, an investigation revealed that a valve had just been repaired and bonnet fasteners were replaced with mismatched and wrong material.[6]
- Each year around 35 million hours are lost because of accidents in United States industries.[16]

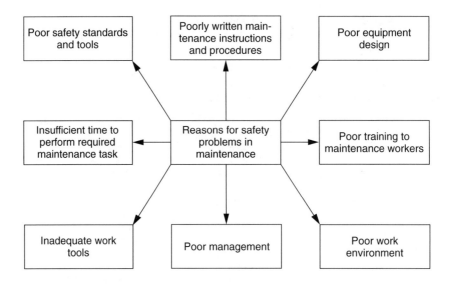

FIGURE 9.2 Some important reasons for safety problems in maintenance.

Accidents occurring during maintenance account for a significant proportion of the overall accidents.

There are various reasons for safety related problems in maintenance. Some of the important reasons are shown in Fig. 9.2.

SAFETY AND MAINTENANCE TASKS

In addition to the general safety considerations, other factors that influence the safety dimensions of maintenance tasks include:

- Numerous maintenance tasks or jobs are in direct response to the needs of working safely. Consequently, safety needs augment maintenance tasks or jobs.
- Numerous maintenance tasks or jobs are hazardous and lead to hazardous solutions. Thus, maintenance work is a cause of safety-related problems.

The first item may be interpreted as one result of an effective safety management system. However, the second item requires further review. Some aspects of maintenance work that give it this dubious safety reputation are as follows:[9]

- Numerous maintenance tasks occur infrequently, e.g., machinery failures, thus fewer opportunities to discern safety-related problems and to introduce remedial measures.
- Maintenance work performed in unfamiliar surroundings means that hazards such as rusted handrails, broken light fittings, and missing gratings may pass unnoticed.

- Difficulty in maintaining regular communication with workers in some maintenance tasks.
- Some maintenance work may require performing tasks such as disassembly of corroded parts, or manhandling cumbersome heavy parts in poorly lit areas and confined spaces.
- Disassembling previously working machinery, thus working under the risk of releasing stored energy.
- Sudden need for maintenance work, allowing limited time to prepare.
- Performance of maintenance work inside or underneath machines such as air ducts, pressure vessels, and large rotating machines.
- Performance of maintenance work at odd hours, in remote locations, and in small numbers.
- Need to transfer heavy and bulky materials from a warehouse to the maintenance workplace, sometimes using lifting and transport equipment well outside a strict maintenance regime.

GUIDELINES FOR EQUIPMENT DESIGNERS TO IMPROVE SAFETY IN MAINTENANCE

One way to improve maintenance safety is to minimize the need for maintenance in systems or equipment at the design stage. If the need for maintenance cannot be eliminated, the designers should follow guidelines for improving maintenance safety as follows:[17]

- Simplify the design as much as possible. Complexity generally adds to maintenance problems.
- Provide fail-safe designs to prevent injury or damage in the event of a failure.
- Develop designs or procedures to minimize the occurrence of maintenance errors.
- Minimize or eliminate the need for special tools/equipment.
- Incorporate devices or other measures to permit early detection or prediction of potential failures so that maintenance can be carried out prior to actual failure with somewhat lower risk of hazard.
- Design for easy accessibility so that items or units requiring maintenance are not difficult to remove, replace, service, or check.
- Develop the design such a way to reduce the probability of maintenance workers being injured by electric shock, contact with a hot surface, escaping high-pressure gas, etc.
- Eliminate the opportunity to perform maintenance or adjustments close to hazardous operating equipment or parts.
- Provide guards against moving parts and interlocks to block access to hazardous locations.
- Consider the typical human behaviors presented in Table 9.2.[18–20]

TABLE 9.2
Typical Human Behaviors

Expected Behavior

Humans tend to use their hands first to test or examine.

Humans usually continue to use faulty items/equipment/systems.

Humans tend to think of manufactured items as being safe.

Humans are easily confused by unfamiliar items.

Humans frequently overestimate bulky weight and underestimate compact weight.

Humans often overestimate speed of an accelerating object and underestimate speed of a decelerating object.

Humans frequently overestimate the probability of occurrence of the "pleasant event" and underestimate the probability of occurrence of the "unpleasant event."

Humans generally know very little about their physical limitations.

MAINTENANCE SAFETY-RELATED QUESTIONS
FOR EQUIPMENT MANUFACTURERS

There are several areas in which attention by equipment manufacturers can improve maintenance safety, directly or indirectly, in the field. In this regard, the following questions can assist manufacturers to determine whether the common problems that might be encountered during the maintenance phase have been properly addressed:[17]

- Are adequate, clear, and easily understandable instructions available for maintenance and repair?
- Were human factors principles used to minimize maintenance problems?
- Can the disassembled item/equipment for repair be reassembled incorrectly so that it becomes hazardous to users?
- Are the items requiring frequent maintenance be as accessible as possible?
- Is there any built-in mechanism which would indicate that safety-critical items require maintenance?
- Is the need for special tools for repairing safety-critical items minimized?
- Do the repair instructions warn when protective gear must be worn because of pending hazards?
- Is it possible to repair the item under consideration by people other than the specially-trained and -equipped personnel?
- Is the repair process hazardous to involved repair workers?
- Are there effective warnings against working on systems that can shock people?
- Is the item designed so that after a failure it would stop operating and not cause damage?
- Is there mechanism to indicate that the redundant units of safety-critical systems have failed?

- Are the voltages reduced to levels at test points to minimize hazards to maintenance workers?
- Are warnings included in instructions to alert repair workers of any danger?
- Is there drive to minimize the cost of safety-critical part replacements?
- Is there an appropriate mechanism to remove fuel or other hazardous fluid from the equipment to be repaired?
- Does the equipment contain safety interlocks that must be bypassed to make adjustments or accomplish repairs?
- Are the test points located at easily reachable locations?

MAINTENANCE PERSONNEL SAFETY

Usually, emphasis is placed on designing safety into machines rather than on the safety of the operators, maintainers, etc. On occasion, more protection is required for maintenance workers beyond the safety designed into machines or processes. Two important areas of maintenance worker safety are respiratory protection and protective clothing. Figure 9.3 shows four areas in which respiratory protection is required. The protective clothing includes items such as:[9]

- *Ear defenders:* These are necessary where machine or process noise can damage maintenance workers' ears.
- *Boots and toecaps:* Well-fitting boots with steel toecaps can reduce the risk of injury in situations such as dismantling used equipment where heavy metal parts are difficult to hold and are likely to slip and drop on exposed feet.
- *Helmets and hard hats:* These are useful to protect maintenance workers from head injury.

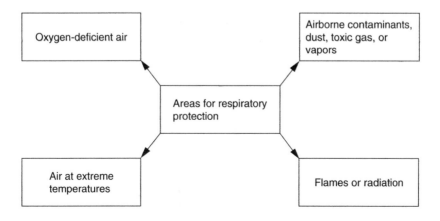

FIGURE 9.3 Areas requiring respiratory protection.

- *Gloves:* These are important to protect hands from injury when performing various types of maintenance tasks.
- *Goggles, visors, screens, and safety glasses:* These items protect eyes from flying chips, sparks, chemical sprays, jetted hydraulic fluid, etc.

PROBLEMS

1. Define the following terms:
   - Maintenance quality assurance
   - Quality
2. Give two examples of engineering system catastrophe due to poor-quality maintenance.
3. Outline the steps to producing good quality maintenance work.
4. A sample of ten machines that underwent maintenance were examined for maintenance-associated defects. Table 9.3 presents a number of maintenance defects found in each machine. Construct the c-chart by using Table 9.3 data and comment on the resulting chart.
5. What are the objectives of post-maintenance testing?
6. What are the important elements of post-maintenance testing?
7. What are the important reasons for the safety problems in maintenance?
8. List at least eight aspects of maintenance work that give maintenance a dubious safety reputation.
9. List at least ten guidelines for equipment designers to observe to improve maintenance safety.
10. List at least seven typical human behaviors designers should consider during equipment design as applicable.
11. List at least fifteen maintenance safety-related questions for equipment manufacturers that can improve maintenance safety of their products.

TABLE 9.3
Maintenance Defects Associated with Each Machine

Machine No.	No. of Maintenance Defects
1	4
2	0
3	3
4	2
5	4
6	8
7	6
8	5
9	2
10	1

REFERENCES

1. Omdahl, T.P., ed., *Reliability, Availability, and Maintainability (RAM) Dictionary,* ASQC Quality Press, Milwaukee, Wisconsin, 1988.
2. McKenna, T. and Oliverson, R., *Glossary of Reliability and Maintenance Terms,* Gulf Publishing Co., Houston, Texas, 1997.
3. Beckmerhagen, I.A., Berg, H.P., and Harnack, K., Quality assurance for safety-related components in a waste repository, *Journal of Quality in Maintenance Engineering,* 2, 1996, 38–49.
4. *Accident Facts,* National Safety Council, Chicago, Illinois, 1999.
5. AMCP 706-132, *Maintenance Engineering Techniques,* U.S. Army Material Command, Department of the Army, Washington, D.C., 1975.
6. *Joint Fleet Maintenance Manual,* Vol. 5, Quality Maintenance, Submarine Maintenance Engineering, U.S. Navy, Portsmouth, NH.
7. Report: The presidential commission on the space shuttle Challenger accident, Vol. I, Washington, D.C., 1986.
8. Elsayed, E.A., *Reliability Engineering,* Addison Wesley Longman, Reading, Massachusetts, 1996.
9. Stoneham, D., *The Maintenance Management and Technology Handbook,* Elsevier Science, Oxford, U.K., 1998.
10. Shewhart, W.A., *Economic Control of Quality of Manufactured Product,* D. Van Nostrand Co., New York, 1931.
11. *Statistical Quality Control Handbook,* Western Electric Co., Indianapolis, Indiana, 1956.
12. Puri, S.C., *Statistical Process Quality Control,* Standards-Quality Management Group, Washington, D.C., 1984.
13. Dhillon, B.S., *Quality Control, Reliability, and Engineering Design,* Marcel Dekker, New York, 1985.
14. DOE-STD-1065-94, *Guideline to Good Practices for Post-maintenance Testing at DOE Nuclear Facilities,* Department of Energy, Washington, D.C., June 1994.
15. *Accident Facts,* National Safety Council, Chicago, Illinois, 1990–1993 edition, 34.
16. Goetsch, D.L., *Occupational Safety and Health,* Prentice-Hall, Englewood Cliffs, New Jersey, 1996.
17. Hammer, W., *Product Safety Management and Engineering,* Prentice-Hall, Englewood Cliffs, New Jersey, 1980.
18. Woodson, W.E., *Human Factors Design Handbook,* McGraw-Hill, New York, 1981.
19. Nertnery, R.J. and Bullock, M.G., Human factors in design, Report No. ERDA-76-45-2, The Energy Research and Development Administration, Department of Energy, Washington, D.C., 1976.
20. Dhillon, B.S., *Advanced Design Concepts for Engineers,* Technomic Publishing Co., Lancaster, Pennsylvania, 1998.

10 Maintenance Costing

INTRODUCTION

The maintenance phase is an important element of the equipment life cycle during which equipment must be maintained satisfactorily for effective performance. The cost of maintaining equipment often varies from 2 to 20 times the acquisition cost. The cost of maintenance is defined as costs that include lost opportunities in uptime, rate, yield, and quality due to nonoperating or unsatisfactorily operating equipment in addition to costs involved with equipment-related degradation of the safety of people, property, and the environment.[1] However, often maintenance cost is simply described as the labor and materials expense needed to maintain equipment/items in satisfactory operational state.[2] References 3 and 4 have classified fundamental costs associated with maintenance more specifically into four areas: direct costs, lost production costs, degradation costs, and standby costs.

Direct costs are associated with keeping the equipment operable and include costs of periodic inspection and preventive maintenance, repair cost, overhaul cost, and servicing cost. Lost production costs are associated with loss of production due to primary equipment breakdown and unavailability of standby equipment. Degradation costs are associated with deterioration in the equipment life due to unsatisfactory/inferior maintenance. Standby costs are associated with operating and maintaining standby equipment. Standby equipment is used when primary facilities are either under maintenance or inoperable.

This chapter presents important aspects of maintenance costing.

REASONS FOR MAINTENANCE COSTING AND FACTORS INFLUENCING MAINTENANCE COSTS

Some of the many reasons for maintenance costing are as follows:

- Determine maintenance cost drivers
- Prepare budget
- Provide input in the design of new equipment/item/system
- Provide input in equipment life cycle cost studies
- Control costs
- Make decisions concerning equipment replacement
- Compare maintenance cost effectiveness to industry averages
- Develop optimum preventive maintenance policies
- Compare competing approaches to maintenance
- Provide feedback to upper level management
- Improve productivity

Many factors influence maintenance costs, including asset condition (i.e., age, type, and condition), operator expertise and experience, company policy, type of service, skills of maintenance personnel, operational environment, equipment specification, and regulatory controls.[5]

MAINTENANCE BUDGET TYPES, PREPARATION APPROACHES, AND STEPS

A maintenance budget serves as an important tool to control financial resources necessary for running the maintenance department. Budget administration uses various types of accounting procedures and computer-based systems to manage, control, and measure departmental effectiveness.[6] One of these two types of budgets is often used in maintenance operations:[7]

- Operating budget
- Project (or appropriation) budget

The operating budget is concerned with itemizing each category of operating expense forecasted for every department in the organization. A purpose of this type of budget is to control normal operating labor, material, and overhead costs forecasted for the coming fiscal year. The budget includes items such as preventive maintenance, semi-annual plant shutdown repairs and overhauls, minor modifications, and routine repair.

The project budget is concerned with special projects or programs such as computerized maintenance management systems, major capital equipment purchases, and major construction projects. Funds for projects such as these are not included in the operating budget. The project budget is divided into particular types and amounts of materials, labor, and overhead expenses needed to complete a defined project.

BUDGET PREPARATION APPROACHES

Two effective approaches that can be used to prepare budgets are discussed here: historical approach and zero-based approach.[7]

With a historical approach, the budget is based on historical perspective. Most budgets fall into this category. Professionals involved in the preparation of this type of budget rely on the experience of earlier years to determine cost estimates for the coming year. This approach is efficient, rational, and requires a relatively small amount of paperwork. The main drawback is that past errors tend to be perpetuated and ineffective operations are funded proportionately to effective ones.

The zero-based approach is concerned with developing the budget from the ground up without any historical basis. Each budget item is justified by current requirements or priority versus the availability of funds. Budget items and subitems are grouped by priority into work packages and, in turn, the work packages are classified into three categories: expenditures required by law, expenditures not required by law, and new or first-time budget items. Some of the advantages of the zero-based budget approach are as follows:

- More thoughtful and thorough process
- Better use of the funds available
- More clear understanding of organizational objectives and goals at all management levels

The zero-based budget approach has drawbacks in that it requires more time because it is a more detailed process, and it requires more documentation in comparison to the historical approach.

MAINTENANCE BUDGET PREPARATION STEPS

Professionals working in the area have developed a process for preparing a maintenance budget. This process is as follows:[7]

- Collect information on trends over the past few years.
- Seek input from the accounting department concerning cost trends and improvements.
- Seek input from the operations group concerning its plans for the coming year.
- Obtain information on sales by product and department.
- Determine maintenance labor-hours by skill and department, particularly for equipment with high repair costs.
- Estimate the amount of material required by department, in particular high-cost and high-volume items.
- Estimate overhead expenses.
- Distribute expenses or costs by weeks and total them for each month.
- Establish separate cumulative cost charts for every important variable, e.g., material, labor, and overhead.
- Update individual and total costs periodically and plot them on appropriate charts.

The last two steps are basically concerned with the budget use.

MAINTENANCE LABOR COST ESTIMATION

The cost of labor is an important component of the maintenance cost. Labor costs are made up of payroll information that is usually obtained from labor distribution reports prepared by the accounting department. The information is required for four key areas:[7]

1. Total number of hours worked annually on a per-employee basis
2. Hourly cost of employee benefits on a per-employee basis
3. Ratio of cost of annual benefits to yearly wages
4. Base pay rates per hour by labor grade

The cost per employee is expressed by

$$C_{em} = LR(1 + BR)TAH \qquad (10.1)$$

where
   C_{em} = cost per employee,
   LR   = hourly labor rate,
   BR   = benefit ratio,
   TAH = total number of annual hours.

The total labor cost is given by

$$TLC = C_{em}N \qquad (10.2)$$

where
   TLC = total labor cost,
   N = number of employees.

Example 10.1

Assume that in a maintenance organization we have the following data:

- TAH = 2000 h
- BR = 0.2
- LR = \$15 per hour
- $N = 40$

Calculating the total labor cost associated with the maintenance activity by inserting the above values into Eqs. (10.1) and (10.2), we get

$$C_{em} = 15 \times (1 + 0.2) \times 2,000 = \$36,000$$

and

$$TLC = 36,000 \times 40 = \$1,440,000$$

Thus, the total labor cost associated with the maintenance activity will be \$1.44 million.

STANDARD HOURLY COST ESTIMATION

The cost per standard hour provides a unit of measure useful for comparing maintenance effectiveness on a consistent basis within the maintenance organization. For example, as maintenance workers produce service, not uniform pieces or parts, the cost per standard hour is a useful tool to compare electrical work, carpentry, pipefitting, etc., on an equitable basis.

The cost per standard hour produced is defined by:[7]

$$C_{sh} = \frac{PD \times FBF}{PHP + EPHP}$$ (10.3)

where
$\quad C_{sh}$ = cost per standard hour produced,
$\quad$ PD $\quad$ = payroll dollars per period,
$\quad$ PHP $\quad$ = planned hours produced,
$\quad$ EPHP = equivalent planned hours produced,
$\quad$ FBF $\quad$ = fringe benefit factor.

The main benefit of the cost per planned hour method over the cost per actual hour method is that it highlights performance-related variations.

MANPOWER REPAIR COST ESTIMATION

This model was developed by the U.S. military to estimate repair costs with respect to manpower.[8] The repair cost is expressed by

$$RC = \alpha(1 - RSF)C_{um}$$ (10.4)

where
$\quad$ RC $\quad$ = repair cost with respect to manpower,
$\quad C_{um}$ $\quad$ = unit repair cost with respect to manpower,
$\quad$ RSF = repairable shrinkage factor due to loss, damage, etc. Its values are tabulated
$\qquad$ in Reference 8 and vary from 0 to 0.1375.
$\quad \alpha$ $\quad$ = number of repairable units failing over the system lifespan.

The number of repairable units failing over the system lifespan is given by

$$\alpha = \lambda n_r L_s H_o$$ (10.5)

where
$\quad L_S$ = system life; in Reference 8 it was taken as ten years,
$\quad H_o$ = operating hours per year,
$\quad n_r$ = total number of repairable items,
$\quad \lambda$ $\;$ = item constant failure rate.

The unit repair cost with respect to manpower is expressed by

$$C_{um} = MCH \times AMHPR \times MUF$$ (10.6)

where
$\quad$ MCH $\quad$ = manpower cost per hour including overhead,
$\quad$ AMHPR = average number of man-hours per repair action,
$\quad$ MUF $\quad$ = manpower use factor and its values are tabulated in Reference 8. The
$\qquad$ tabulated values vary from 3 to 1.04.

CORRECTIVE MAINTENANCE LABOR COST ESTIMATION

In this case corrective maintenance labor cost is estimated when the item/system mean time between failures (MTBF) and mean time to repair (MTTR) are known. Consequently, the annual labor cost of corrective maintenance is expressed by[2]

$$CM_{al} = \frac{SOH \times LCH \times MTTR}{MTBF} \tag{10.7}$$

where
   CM_{al} = annual corrective maintenance labor cost,
   LCH = corrective maintenance labor cost per hour,
   SOH = annual scheduled operating hours.

Example 10.2

A system is scheduled to operate for 2000 hours per year. The system's MTBF and MTTR are 400 h and 20 h, respectively. Calculate the annual labor cost of corrective maintenance if the maintenance labor cost is $20 per hour.
   By substituting the specified data into Eq. (10.7), we get

$$CM_{al} = \frac{2000 \times 20 \times 20}{400} = \$2000$$

The annual labor cost of corrective maintenance is $2000.

MAINTENANCE MATERIAL COST ESTIMATION

The cost of maintenance materials is an important component of the total maintenance cost. In U.S. industry, maintenance materials typically account for 40 to 50% of the total maintenance cost.[3] Well-planned and efficiently operated stock and storerooms can significantly reduce the cost of materials as the costs of excessive inventory and obsolete parts are important factors in most maintenance storerooms and stockrooms. During the costing of store items used in maintenance work, factors such as those listed below should be considered.

   • Cost associated with inventorying the material
   • Latest procurement or manufacturing cost
   • Cost associated with the invested capital
   • Reduction in stock item value due to decay or spoilage
   • Increase in stock item value due to inflation

   The total cost of stock or stores at the time of repair is given by[3]

$$TCS = PDC + IC + (WI - PDC) + (0.01 \times T \times PDC) + (0.1 \times PDC)$$

$$= WI + IC + \left[\frac{(T \times PDC) + (10 \times PDC)}{100} \right] \tag{10.8}$$

where
   TCS = total cost of stock or stores at the time of repair,
   PDC = present value of the inventory item cost including purchase price and
           delivery cost,
   WI   = worth of the inventory item after n periods,
   IC   = inventory cost per item,
   T = time, expressed in months, the stock item is in inventory.

Note that Eq. (10.8) allows a rate of inflation of 1% per month of purchase cost while the item is in inventory, and a 10% allowance for the item's entire shelf life to take into consideration deterioration, spoilage, obsolescence, and theft.
   Equations for calculating PDC, WI, and IC are given below, respectively.

$$\text{PDC} = w \times \text{PP} \times (1 + L_S + L_u) - \text{SM} \tag{10.9}$$

$$\text{WI} = \text{PDC} \times (1 + i)^n \tag{10.10}$$

$$\text{IC} = \frac{\text{FSC} \times B}{K \times R} \tag{10.11}$$

where
   PP   = purchase price of material per unit; more specifically, the delivered price,
   L_u = losses generated by the unused stock returned to inventory that is too
           small for future use,
   w = weight/other unit of quantity of material used,
   SM   = unit price of material salvaged,
   L_S = losses due to scrap, skeletons, chips, etc.,
   i = interest rate for a specified period,
   n = number of interest periods,
   FSC = annual floor space cost per square foot,
   B = bin size expressed in square feet,
   R = reciprocal of years item normally spends in inventory,
   K = average number of items stored in bin.

MAINTENANCE COST ESTIMATION MODELS

This section presents models to estimate costs other than specific maintenance labor and materials. This section also presents examples of models used to estimate the maintenance cost of specific items.

BUILDING COST ESTIMATION MODEL

The building cost (BC) is expressed by:[8]

$$\text{BC} = \text{NCF} \times \text{ICPCF} \tag{10.12}$$

where

    ICPCF = initial cost per cubic foot,

    NCF   = number of cubic feet needed for maintenance buildings.

MAINTENANCE EQUIPMENT COST ESTIMATION MODEL

The cost of maintenance equipment can be estimated using the following equation:[8]

$$CME = RDC + \theta(UPC) \tag{10.13}$$

where

    CME = maintenance equipment cost,

    RDC = research and development cost associated with the maintenance equipment,

    θ = total number of maintenance equipment,

    UPC = maintenance equipment unit procurement cost.

PRODUCTION FACILITY DOWNTIME COST ESTIMATION MODEL

A production facility downtime cost (PFDC) is expressed by[5]

$$PFDC = S_i + C_r + RC_r + RL + C_c + C_p \tag{10.14}$$

where

    S_i = salary of idle operator,

    C_r = rental cost of replacement unit (if any),

    RC_r = cost for replacement of ruined product,

    RL   = revenue loss, less recoverable costs like materials,

    C_c = tangible and intangible costs associated with customer dissatisfaction, loss of goodwill, hidden costs, etc.,

    C_p = costs associated with late penalties, etc.

AVIONICS COMPUTER MAINTENANCE COST ESTIMATION MODEL

The maintenance cost of an avionics computer is expressed by[9]

$$C_{mc} = C_{am} \, X/1000 \tag{10.15}$$

where

    C_{mc} = total cost of computer maintenance,

    C_{am} = annual maintenance cost per unit expressed in 1974 dollars ($\times 10^3$),

    X = total number of years in operation.

The natural logarithm of C_{am} is expressed by

$$\ln C_{am} = \alpha_1 + \alpha_2 \ln C_{un} - \alpha_3 \ln(MTBF) \tag{10.16}$$

where
　　MTBF = mean time between failures expressed in hours,
　　C_{un}　　= unit cost expressed in 1974 dollars ($\times 10^3$),
　　α_1　　= 6.944,
　　x_2　　= 0.296,
　　x_3　　= −0.63.

FIRE CONTROL RADAR MAINTENANCE COST ESTIMATION MODEL

The fire control radar maintenance cost is defined by[10]

$$C_{fr} = C_{mh}\, H_y X / 1000 \tag{10.17}$$

where
　　C_{fr}　= total fire control radar maintenance cost,
　　C_{mh} = maintenance cost per flying hour per unit expressed in 1974 dollars
　　　　($\times 10^3$),
　　X　 = total number of years in operation,
　　H_y　= number of flying hours per year.

The natural logarithm of C_{mh} is

$$\ln C_{mh} = \beta_1 + \beta_2 \ln P_k \tag{10.18}$$

where
　　P_k = peak power in kilowatts,
　　β_1 = −2.086,
　　β_2 = 0.611.

DOPPLER RADAR MAINTENANCE COST ESTIMATION MODEL

The maintenance cost of a Doppler radar is expressed by[10]

$$C_{dr} = C_{drm} X / 1000 \tag{10.19}$$

where
　　C_{dr}　 = total Doppler radar maintenance cost,
　　C_{drm} = annual maintenance cost of the Doppler radar,
　　X　　 = total number of years in service.

The natural logarithm of C_{drm} is

$$\ln C_{drm} = \mu_1 + \mu_2 \ln C_f \tag{10.20}$$

where

C_f = Doppler radar first unit cost expressed in 1974 dollars ($\times 10^3$),
$\mu_1 = -1.269$,
$\mu_2 = 0.696$.

EQUIPMENT OWNERSHIP CYCLE MAINTENANCE COST ESTIMATION

This section is concerned with estimating the maintenance cost of the entire ownership cycle of equipment, so that its present value can be added to acquisition cost to obtain equipment life cycle cost. The following two formulas are often used to find present value of a sum of money. They can equally be used to obtain present values of equipment ownership cycle maintenance costs:[11]

Formula I

This formula is used to estimate the present value of a single amount of money after k periods and is expressed by

$$PV = \frac{AM}{(1+i)^k} \qquad (10.21)$$

where
PV = present value,
AM = single amount,
k = number of interest or conversion periods (normally taken as years),
i = interest rate per period.

Formula II

This formula is concerned with obtaining the present value of equal amounts of, say, maintenance costs occurring at the end of each of k conversion periods (usually years). The present value is given by

$$PV = MC\left[\frac{1-(1+i)^{-k}}{i}\right] \qquad (10.22)$$

where
MC = maintenance cost occurring at the end of each (interest) conversion period.

Example 10.3

A maintenance department considers procuring an engineering system. Two manufacturers are bidding to provide the system and their corrective maintenance cost-related data are given in Table 10.1.

Determine which of the two systems will be less costly with respect to present values of corrective maintenance and by how much?

TABLE 10.1
Maintenance Cost-Related Data

Description	Manufacturer A System	Manufacturer B System
Expected life	12 years	12 years
Expected cost of a corrective maintenance action	$1000	$1400
Annual failure rate	2.5 failures per year	2 failures per year
Annual interest rate	8%	8%

Manufacturer A System

The annual expected cost, AC_{cm}, of a corrective maintenance action is

$$AC_{cm} = 1000 \times 2.5 = \$2500$$

The present value, PV_A, of the engineering system life cycle corrective maintenance cost using Eq. (10.22) and the specified and calculated values is

$$PV_A = 2500\left[\frac{1 - (1 + 0.08)^{-12}}{0.08}\right]$$

$$= \$18,840.20$$

Manufacturer B System

The annual expected cost, AC_{cm}, of a corrective maintenance action is

$$AC_{cm} = 1400 \times 2 = \$2800$$

The present value, PV_B, of the engineering system life cycle corrective maintenance cost using Eq. (10.22) and the specified and calculated values is

$$PV_B = 2800\left[\frac{1 - (1 + 0.08)^{-12}}{0.08}\right]$$

$$= \$21,101.01$$

Manufacturer A's engineering system will be less costly with respect to present values of corrective maintenance by $2,260.81.

MAINTENANCE COST-RELATED INDICES

Many indices have been developed to measure the effectiveness of maintenance activity with respect to cost. Usually, the values of these indices are plotted periodically to monitor trends. Some of the cost-related indices are presented below.[3,12,13]

MAINTENANCE COST RATIO

This is defined by

$$R_{mc} = \frac{TMC}{TCS} \qquad (10.23)$$

where

R_{mc} = maintenance cost ratio,
TMC = total maintenance cost,
TCS = total cost of sales.

The value of this index averages around 5% for many industries.[13] This index is a maintenance figure-of-merit to compare maintenance costs to all costs.

MAINTENANCE LABOR COST TO MATERIAL COST RATIO

This is expressed by

$$R_{\ell m} = \frac{TMC_\ell}{TMC_m} \qquad (10.24)$$

where

$R_{\ell m}$ = maintenance labor cost to material cost ratio,
TMC_ℓ = total maintenance cost associated with labor,
TMC_m = total maintenance cost associated with materials.

MAINTENANCE COST TO TOTAL OUTPUT RATIO

This is given by

$$R_{mo} = \frac{TMC}{TO} \qquad (10.25)$$

where

R_{mo} = ratio of total maintenance cost to total output,
TO = total output usually expressed in units such as megawatts, tons, and gallons.

This ratio may also be described as a maintenance figure-of-merit for capital manufacturing equipment.

MAINTENANCE COST TO TOTAL MANUFACTURING COST RATIO

This is defined by

$$R_{mm} = \frac{TMC}{C_{tm}} \times 100 \qquad (10.26)$$

where

R_{mm} = ratio (percentage) of total maintenance cost to total manufacturing cost,
C_{tm} = total manufacturing cost.

MAINTENANCE COST TO VALUE OF FACILITY RATIO

This is expressed by

$$R_{mv} = \frac{TMC}{TCI} \tag{10.27}$$

where

R_{mv} = ratio of total maintenance cost to value of facility,
TCI = total cost of investment in plant and equipment.

MAINTENANCE COST TO TOTAL MAN-HOURS WORKED RATIO

This is given by

$$R_{mh} = \frac{TMC}{TMH} \tag{10.28}$$

where

R_{mh} = ratio of total maintenance cost to total man-hours worked, specifically,
          the cost of a maintenance hour in dollars,
TMH = total man-hours worked.

PREVENTIVE MAINTENANCE COST TO TOTAL BREAKDOWN COST RATIO

This is defined by

$$R_{mb} = \frac{TPMC(1000)}{TBC} \tag{10.29}$$

where

R_{mb} = ratio (percentage) of total preventive maintenance cost to total break-
          down cost,
TPMC = total preventive maintenance cost including production losses,
TBC   = total breakdown cost.

COST DATA COLLECTION

In maintenance costing, various types of cost data are required. Management decides the type of cost data the maintenance section or department should collect by keeping in mind its future applications. This section briefly discusses the collection of the following types of cost data:[14]

- *Labor Costs:* Generally, the timesheet is used to obtain data on labor costs. Although the timesheet is a useful source of labor costs for maintenance and accounting divisions, it is insufficient to obtain data on the total cost of individual work orders. Under such circumstances, additional data are needed to determine the costs by specific category of work, skill, or job. In such cases, it is more appropriate to use a job ticket or a work order that can be designed to make readily available all hours recorded for specific maintenance jobs.

- *Equipment costs:* These costs are obtained from either the supplier's invoice or the purchase order.
- *Costs of spare parts and supplies:* In maintenance work such costs are more difficult to allocate than equipment costs. For example, it is not practically feasible to charge individually for items such as nuts and bolts. Nonetheless, a work order is an important source for obtaining this cost data.
- *Overhead Costs:* Normally, these costs are obtained from the accounting department. The maintenance department should examine their accuracy from time to time.

PROBLEMS

1. List at least ten reasons for maintenance costing.
2. What are the important factors that influence maintenance cost?
3. Outline steps used to prepare a maintenance budget.
4. Describe these two types of maintenance budgeting approaches:
   - Historical approach
   - Zero-based approach
5. Assume that a machine is scheduled to operate for 3000 h annually. The machine mean time between failures and mean time to repair are 600 h and 15 h, respectively. Calculate the annual labor cost of corrective maintenance if the labor cost is $35 per hour.
6. Prove that the present value of a single amount of money after n periods is given by

$$PV = \frac{Z}{(1+i)^n} \tag{10.30}$$

where
   PV = present value,
   Z = single amount of money,
   n = number of interest or conversion periods,
   i = interest rate per period.

7. Discuss sources for the following types of maintenance cost-related data:
   - Labor costs
   - Overhead costs
8. Discuss two indices related to maintenance costing.

TABLE 10.2
Maintenance Cost-Related Data

Description	Manufacturer A Machine	Manufacturer B Machine	Manufacturer C Machine
Expected life (years)	10	10	10
Annual failure rate	1.5 failures per year	1 failure per year	2 failures per year
Expected cost of a corrective maintenance action	$1200	$1400	$1100
Annual interest rate	10%	10%	10%

9. Discuss these two types of maintenance budgets:
   - Operating budget
   - Project budget
10. An organization is considering buying a machine. Three manufacturers are bidding, and their corrective maintenance cost-related data are given in Table 10.2. Determine which machine will be most cost-effective with respect to present values of corrective maintenance.

REFERENCES

1. McKenna, T. and Oliverson, R., *Glossary of Reliability and Maintenance Terms,* Gulf Publishing Co., Houston, Texas, 1997.
2. Dhillon, B.S., *Life Cycle Costing,* Gordon and Breach Science Publishers, New York, 1989.
3. Niebel, B.W., *Engineering Maintenance Management,* Marcel Dekker, New York, 1994.
4. Cavalier, M.P. and Knapp, G.M., Reducing preventive maintenance cost error caused by uncertainty, *Journal of Quality in Maintenance Engineering,* 2, No. 3, 1996, 21–36.
5. Levitt, J., *The Handbook of Maintenance Management,* Industrial Press, New York, 1997.
6. Hartmann, E., Knapp, D.J., Johnstone, J.J., and Ward, K.G., *How to Manage Maintenance,* American Management Association, New York, 1981.
7. Westerkamp, T.A., *Maintenance Manager's Standard Manual,* Prentice-Hall, Paramus, New Jersey, 1997.
8. AMCP 706-133, *Engineering Design Handbook: Maintainability Engineering Theory and Practice,* Department of the Army, Department of Defense, Washington, D.C., 1976.
9. Earles, M.E., *Factors, Formulas, and Structures for Life Cycle Costing,* Eddins-Earles, Concord, Massachusetts, 1978.
10. *Cost Analysis of Avionics Equipment,* Vol. I, Prepared by the Air Force Systems Command, Wright-Patterson Air Force Base, Ohio, February 1974. The NTIS Report No. AD 781132. Available from the National Technical Information Service (NTIS), Springfield, Virginia.

11. Dhillon, B.S., *Engineering Design: A Modern Approach,* Richard D. Irwin, Chicago, Illinois, 1996.

12. ER HQ-0004, *Maintenance Manager's Guide,* Energy Research and Development Administration, Washington, D.C., 1976.

13. Omdahl, T.P., ed., *Reliability, Availability, and Maintainability (RAM) Dictionary,* ASQC Quality Press, Milwaukee, Wisconsin, 1988.

14. Jordan, J.K., *Maintenance Management,* American Water Works Association, Denver, Colorado, 1990.

11 Software Maintenance

INTRODUCTION

Today computers find applications in virtually all areas of life. Over the years, there has been a shift in money spent on developing computer from hardware to software. For example, in 1955, the software component including software maintenance accounted for 20% of the total computer cost and three decades later in 1985, the figure increased to 90%.[1] In the early 1950s, there were 100-line programs, and in the latter years of the 1990s they increased to multimillion-line programs.[2]

During the early years of computing, software maintenance was a relatively small component of the overall software life cycle. Over the years, it has become a major factor. For example, in 1955, the proportion of time spent on maintenance was 23%; in 1970, it was 36%; and the prediction for 1985 was 58%.[3,4] Software maintenance may be defined as the process of modifying the software system/component subsequent to delivery to rectify faults, improve performance or other attributes, or adapt to a change in the use environment.[5,6]

Today, organizations around the world spend a vast amount of money on software maintenance. For example, in 1983 the U.S. Department of Defense spent $2 billion on software maintenance, while in the mid-1980s the figure for the entire country was approximately $30 billion per year.[7]

Software maintenance is an important component of the software life cycle and inability to undertake it efficiently, cheaply, and safely can lead to many problems. This chapter presents important aspects of software maintenance.

SOFTWARE MAINTENANCE FACTS AND FIGURES

The following facts and figures are associated with software maintenance:

- Software maintenance accounts for between 40 and 90% of total life cycle costs.[8]
- A study by Hewlett-Packard revealed that 60 to 80% of its research and development staff are involved in maintenance of existing software.[9]
- In business programming, COBOL is the most commonly used language and is not designed for easy maintenance.[10]
- Over two-fifths of software maintenance activities are due to modifications and extensions requested by the users.[11]
- In the early 1990s the U.S. Department of Defense spent approximately $30 billion per year on software. About two-thirds of that was devoted to sustaining deployed software systems.[12]

- A software product's typical life span is 1 to 3 years in development and 5 to 15 years in use (maintenance).[13]
- Over 80% of a software product's life is spent in maintenance.[14]
- Annually for every one thousand delivered source instructions, 92 or 9.2% of them are changed.[15]
- For every year of software system in production, the programming time expended to maintain it adds up to 17 to 33% of the original programming time.[16]
- When a software system costing $1 million over its life span is retired or taken out of service, only less than 5% of it will have changed, but the cost associated with changing that 5% will be at minimum as much as the cost of producing the 95% that was not changed.[17]
- An average software production program will be maintained by ten individuals prior to its being rewritten or discarded.[16]
- A study conducted by the Boeing Company revealed that on average 15% of the lines of source code in a simple or easy program are changed every year, 5% in medium programs, and 1% in difficult programs.[18]
- For every dollar spent on developing software, another dollar must be budgeted to keep that software viable over its life span. Furthermore, another dollar can be spent to make desirable changes.[19]
- For all software systems combined, the maintenance component of the overall total effort is increasing around 3% per year.[4]
- Most software maintenance is development in disguise, of which about 20% is correction of errors.[4,20]
- The maintenance of existing software can consume over 60% of all efforts associated with development.[11]
- Software maintenance accounts for approximately 70% of the software cost.[11]
- The U.S. government spends approximately 40% of the total software cost on maintenance.[21]

SOFTWARE MAINTENANCE IMPORTANCE, EFFORT DISTRIBUTION, AND REQUEST TYPES

The importance of software maintenance varies from one organization to another. A study of attitudes about the relative importance of new development vs. maintenance in an organization is presented in Reference 15. Approximately 55% of the respondents stated that maintenance is more important than new development and, in fact, around 90% indicated that the maintenance is at least equal in importance to new development.

Importantly, the study noted that the higher the individuals in management hierarchy, the more they valued maintenance over new development.[10]

The distribution of maintenance effort in 487 organizations surveyed may be classified as follows:[10,15]

- Enhancements for users: 41.8%
- Data environment adaptation: 17.3%

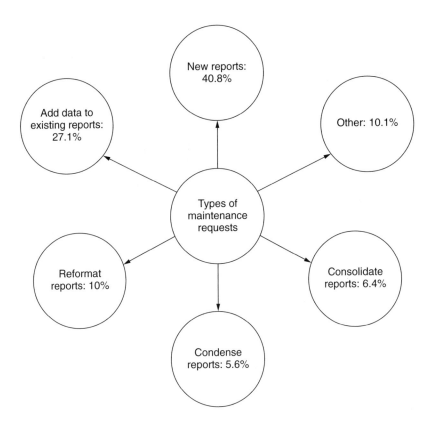

FIGURE 11.1 Maintenance request types.

- Emergency debugging: 12.4%
- Routine debugging: 9.3%
- Changes made to hardware: 6.2%
- Improvements to documentation: 5.5%
- Improvements to code efficiency: 4%
- Other: 3.4%

The maintenance requests in organizations surveyed can be classified into six categories,[10,15] shown in Fig. 11.1. The figure also shows the percentage distribution of these requests.

TYPES OF SOFTWARE MAINTENANCE

Software maintenance focuses on four aspects of system evolution simultaneously: maintaining control over the day-to-day operations of the system, maintaining control over modifications associated with the system, perfecting existing and acceptable functions, and preventing degradation of system performance to unacceptable levels. These four activities are also known as corrective maintenance, adaptive maintenance, perfective maintenance, and preventive maintenance, respectively.[22]

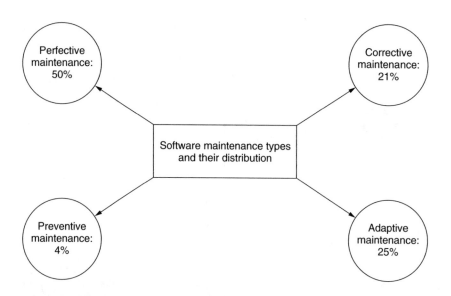

FIGURE 11.2 Software maintenance types and their distribution.

A survey of 487 software development organizations revealed the percentage distribution of the above four types of maintenance, shown in Fig. 11.2.[23] Each of these maintenance types is described below.

1. *Corrective maintenance:* This process incorporates diagnosis and rectification of errors. To control the day-to-day system functions maintenance people respond to problems arising from faults.

   Some ways to improve corrective maintenance are to employ high-level languages, use structured techniques, and keep modules as small as possible.[24]

2. *Adaptive maintenance:* This activity modifies software to effectively interface with a changing environment (i.e., both hardware and software). Note that the adaptive changes made to add parameters do not rectify faults, they only permit the system to adapt appropriately as it evolves.

   Striving for hardware independence and using a portable high-level language improve adaptive maintenance.[24]

3. *Perfective maintenance:* This activity adds capabilities, modifies existing functions, and makes general enhancements. Perfective maintenance involves making modifications to enhance some aspect of the system, even when such modifications are not dictated by faults. Perfective maintenance consumes around 50% of the total maintenance effort.

   Improving only modules with a high degree of usage, a reasonable life span, and a high cost for performing adaptive or corrective maintenance can be quite useful in perfective maintenance.[24]

4. *Preventive maintenance:* This activity modifies software to enhance potential reliability/maintainability, or provides an improved basis for future enhancements. Usually, this type of maintenance is practiced when involved software

professionals discover an actual or potential fault that has not yet become a failure and take appropriate corrective measures. Preventive maintenance is still relatively rare.

SOFTWARE MAINTENANCE PROBLEMS

The maintenance of software systems is difficult because they are already operational. Thus, it is necessary to keep appropriate balance between the need for change and keeping the system accessible to users. Many technical and managerial problems result when changing software quickly and cheaply. Furthermore, as many software systems under maintenance are fairly large and complex, the solutions which may work well for laboratory-scale pilots but do not scale up to industrial or real life-sized software.

Various authors have grouped software maintenance problems differently. For example, Reference 2 has grouped them into three categories: (1) alignment with organizational objectives, (2) process issues, (3) technical issues. On the other hand Reference 22 has classified them into areas such as staff problems, the need to compromise, maintenance cost, and technical problems.

Some of the people-related software maintenance problems include poor understanding of the maintenance needs and low morale of the professionals involved. A clear understanding of what needs to be changed is important because around 47% of the software maintenance effort is associated with comprehending the software to be modified.[25] This high figure results from the number of interfaces that need to be examined in the event of changing a component. Also, more than 50% of maintenance professionals' problems are due to users' poor understanding or skills.[23]

An important factor for the low morale is the second-class status often accorded to maintenance persons. The results of a study indicate that approximately 12% of the problems during maintenance are the result of low morale and productivity.[23]

With respect to technical maintenance problems, a change made at one place in the software system may have a ripple effect elsewhere. This means that understanding the consequences of changes is essential. For a change is to be consistent, maintenance persons must investigate the possibility of occurrence of all types of ripple effects. Ripple effect propagation may be defined as a phenomenon by which modifications made to a software element during the software life cycle (i.e., specification, design, code, or test phase) affects other elements or components.[2]

SOFTWARE MAINTAINABILITY

Software maintainability may be viewed in two different ways: (a) reflecting the external view of the software, (b) reflecting the internal view of the software. The reason for the first way is that maintainability depends on the product itself, as well as on individuals involved in maintenance, proposed software usage, and supporting documentation and tools.[22] In short, it is impossible to measure maintainability without monitoring the behavior of software in a specific environment.

The measurement of maintainability prior to actually delivering the software is considered important due to various factors, including getting a sense of the resources required to support any problems that may occur. Under such circumstances the internal software attributes, i.e., those relating to the structure, are used.

Maintainability is not only restricted to code. It also describes software products such as specification, design, and test plan documents. Consequently, maintainability measures are necessary for all products to be maintained.

Broadly speaking, the maintainability of software is a qualitative measure of factors such as ease of understanding (i.e., the structure, interfaces, functions, and internal procedure of the software), ease of diagnosis and testing, and ease of change.[24]

EXTERNAL VIEW

The external view of software maintainability includes measures such as:

- Mean time to repair
- Total number of unresolved problems
- Total amount of time spent on resolved problems
- Total number of components modified for implementing a change
- Ratio of total change implementation time to the number of changes implemented
- Percentage of modifications that introduced new faults in software system

Probably the most effective maintainability measure is mean time to repair. For its calculation, the careful records of information such as those listed below are required.[22]

- Problem reporting time
- Time needed to perform analysis of the problem
- Total time lost due to administrative delay
- Total amount of time needed to specify which modifications are to be made
- Total amount of time required to make the change
- Total amount of time required to document the change
- Total amount of time required to test the change

INTERNAL VIEW

There are various measures for internal attributes of software relating to maintainability proposed by many researchers. For example, complexity measures often correlate with the maintenance effort, i.e., the more complex the code, the greater the effort required for maintenance. Although correlation and measurement are not the same, there is a relationship between poorly structured and inadequately documented products and their maintainability. Some of the measures are presented below.

Cyclomatic Number

This number was first defined by T. McCabe in 1976, and is one of the most frequently used measures during maintenance.[26] The cyclomatic number is a metric that takes into consideration the structural complexity of source code by measuring the number of linearly independent paths in the code. It is based on graph-theoretic concepts, and is computed by converting the code into its equivalent control flow graph and then applying graph properties. In maintenance work, the number of independent paths indicates the degree to which it is required to comprehend and track when examining or changing a component. Per Lehman's second law of software evolution, the cyclomatic number and other complexity measures will increase as the software system evolves. The McCabe's complexity number is expressed by:[26-28]

$$CN_M = E - n + 2y \qquad (11.1)$$

where

CN_M = McCabe's complexity number,
E = total number of edges in the software program under study,
n = total number of nodes or vertices,
y = total number of connected components or separate tasks.

Example 11.1

Assume $n = 9$, $E = 6$, and $y = 2$. Calculate the value of the McCabe's complexity number.

Substituting the specified values into Eq. (11.1), we get $CN_M = 6 - 9 + 2 \times 2 = 1$. This result indicates that high reliability of the software under consideration can be expected. The higher value of the complexity number means that it will be more difficult to maintain software.

Fog Index

Readability affects maintainability, particularly for textual products. The Fog index is a useful readability measure and is defined by:[22,29-30]

$$I_f = (0.4)\left[\frac{n_w}{n_s} + P\right] \qquad (11.2)$$

where

I_f = Fog index,
P = percentage of words of three or more syllables,
n_s = total number of sentences,
n_w = total number of words.

This measure is purported to correspond approximately with the years of education an individual will require to understand a passage with ease.

Source Code Readability Index

This measure is specifically designed for software products and is defined by[31]

$$R_{SC} = 0.295X - 0.499Y + 0.13Z \qquad (11.3)$$

where

R_{SC} = source code readability,

X = mean normalized length of variables (a variable's length is the number of characters in a variable name),

Y = total number of lines containing statements,

Z = McCabe's complexity or cyclomatic number.

SOFTWARE MAINTENANCE TOOLS AND TECHNIQUES

Over the years many tools and techniques have been developed that directly or indirectly concern software maintenance. A survey of software maintenance tools is presented in Reference 32. This section describes some of the tools and techniques in detail.[4,22]

SOFTWARE CONFIGURATION MANAGEMENT

Software configuration management can be defined as a set of tracking and control activities that starts at the beginning of a software development project and terminates at the retirement of the software. Keeping track of changes and their effects on other system components or parts is a challenging task. Usually, the more complex and sophisticated the system under consideration, the more parts or components a change will affect. For this reason, configuration management is an important and critical factor during maintenance.

Configuration management is practiced by establishing a configuration control board because many maintenance-related changes are requested by users or customers to correct failures or make enhancements. The board oversees the change process, and its membership includes interested parties: customers, users, and developers. Each highlighted problem is handled in the following manner:[22]

1. A user, developer, or customer who discovers a problem uses a formal change control form to record all associated symptoms. Similarly, in the case of enhancement, all relevant information is recorded.
2. The configuration control board is formally informed of the proposed change.
3. The board meets and discusses the proposed change.
4. After making a decision about the change requested, the board prioritizes the change and assigns appropriate individuals to make the change.
5. The designated individual(s) identifies the problem source and the highlights the changes required. Working with the test copy, the assigned individual tests and implements the changes.
6. The designated individual(s) works with the software program librarian to control and track the change or modification installation in the operational system and update associated documentation.
7. A change report describing the changes made is filed by the designated individual.

Step 6 above is the most critical step because at any time the configuration management team must be aware of the status of any component/document in the system. This requires effective communication among involved individuals. Consequently, it is necessary to have answers to questions listed below.[33]

- What was changed?
- When did the change occur?
- Who is responsible for the change under consideration?
- Who authorized the change?
- Who was made aware of the change?
- Who actually made the change?
- What is the change priority?
- Who can terminate the change request?
- Was the change made effectively and correctly?

The above questions take into consideration naming, synchronization, delegation, authorization, routing, identification, valuation, cancellation, and authentication, respectively.

IMPACT ANALYSIS

Software maintenance depends on and starts with customer or user requirements. A requirement translating into a seemingly minor change is frequently more extensive; consequently, more costly to implement than anticipated. Under such circumstances a study of the impact of the change could provide useful information, especially where the change is complex and sophisticated.

Impact analysis can be defined as the determination of risks associated with the proposed change, including the estimation of effects on factors such as effort, resources, and schedule. Several ways to measure the impact of a change are given in Reference 34.

MAINTENANCE REDUCTION

Reduction in the amount of maintenance helps increase maintenance productivity. A maintenance staff armed with the latest knowledge, skills, and techniques can reap significant productivity and quality improvements.

Some methods for reducing software maintenance are as follows:[4,13,20,35–41]

- Use of portable languages, operating systems, and tools.
- Use preventive maintenance approaches, such as using limits for tables that are reasonably greater than possibly be required.
- Highlight possible enhancements and design the software so that it can easily incorporate those enhancements.
- Consider human factors in areas such as screen layouts during software design. This is one source of frequent changes or modifications.
- Introduce structured maintenance that employs approaches for documenting currently existing systems and includes guidelines for reading programs, etc.

- Divide the functions into two groups: most likely to be changed and inherently more stable.
- Employ standard methodologies.
- Store constants in tables instead of scattering them throughout the program.
- Encourage effective communication among maintenance programmers.
- Schedule maintenance on specific dates only and do not allow changes in between those dates.

AUTOMATED TOOLS

In software work, tracking the status of all components and tests is challenging. Fortunately, there are automated tools available on the market that are useful in maintaining software. Some of these tools are described below:[22]

- *File Comparator:* This maintenance tool compares two files and determines their differences. File comparators are frequently used to determine if two supposedly identical systems or programs are in fact identical.
- *Text Editors:* These editors are useful, for example, in preventing errors during text duplication, because a text editor can copy documentation or code from one place to another.
- *Static Code Analyzers:* These compute information concerning the code structural attributes, for example, number of lines of code, depth of nesting, cyclomatic number, and number of spanning paths. The information is useful in determining if the new versions of systems are becoming more complex, bigger, and more difficult to maintain.
- *Cross-Reference Generators:* Generators are useful in assuring that the changed code will still comply with its specifications, especially when these generators possess a set of logical formulas known as verification conditions. For example, if all concerned formulas yield a value of "true," the code meets the specifications that produced it.
- *Compilers and Linkers:* These automated tools often possess features that simplify maintenance and configuration management. In the case of a compiler, it checks code for syntax errors and points out the location and fault type. Once the code is compiled correctly, the linker or the link editor establishes links between the code and other components required for operating the program. Some linkers can eliminate problems caused by using incorrect copy of a system/subsystem when testing change.
- *Debugging Tools:* These tools are useful in tracing the program logic step by step, setting flags and pointers, and examining register and memory area contents.

SOFTWARE MAINTENANCE COSTING

Today the cost of maintaining software systems over their life cycles has become an important factor. For example, in the 1970s development consumed most of a software system's total budget. In the 1990s some estimates suggested that maintenance costs may have increased to as high as 80% of a system's life cycle cost.[22]

TABLE 11.1
Factors Affecting Software Maintenance Cost

Factor	Comment
Programming style	It is easier to maintain a well-structured program.
Programming language	It is easier to maintain a high-level language program.
Program age	Usually old programs are more expensive to maintain.
Module independence	A change made to one module that affects others is usually more expensive to maintain.
Staff stability	The stability of the staff involved helps reduce maintenance costs.
Documentation quality	Good quality documentation is easier to understand.
Configuration management	Good configuration management helps maintain links between programs and their documentation.
Hardware stability	Software programs designed for stable hardware will need change with a change in hardware.
Application domain	Usually, it is easier to maintain mature, well-understood application domains.
Program validation and testing	Well-validated software programs generally need fewer changes due to corrective maintenance.
External environment	Software programs that depend on their external environment may have to be changed when the environment changes.

In general the following may be said about software maintenance costs:

- Normally higher than development costs.
- Increase in software is maintained because maintenance corrupts the software structure, thus making further maintenance cumbersome.
- Usually, aging software has high support costs because of old languages, compilers, etc.
- Both technical and nontechnical factors affect maintenance costs significantly.

Maintenance costs are affected by many factors, as listed in Table 11.1.

Over the years various mathematical models directly or indirectly concerned with software maintenance costing have been developed. Three such models are presented below.

MILLS MODEL

This model demonstrates how maintenance cost can build up alarmingly fast. The following assumptions are associated with this model:[27,42]

- Programming work force is constant and is normalized to be unity.
- After project completion, a (normalized) maintenance force, say m, is assigned to do maintenance. Consequently, a (normalized) work force, d, is left for developing new projects.

Thus, we have

$$PF = m(t) + d(t) \qquad (11.4)$$

where
  PF = programming work force,
  t = time.

From Eq. (11.4), we note that at $t = 0$, $m = 0$ and $PF = d = 1$.
The fraction of the development force, y, assigned to maintenance at the completion of a project is defined by

$$y = \frac{m}{PF} \qquad (11.5)$$

But at $t = 0$, y is not defined.

If we start our first project at zero time, then at its release, i.e., $t = t_1$, we have

$$m(t_1) = yd(t_1) = y \cdot 1 = y \qquad (11.6)$$

and

$$d = 1 - m = 1 - y \qquad (11.7)$$

After the release of the second project,

  m = assignment to project no. 1 + assignment to project no. 2
$$= y + yd$$
$$= y + y(1 - y) \qquad (11.8)$$

and

$$d = PF - m$$
$$= 1 - m$$
$$= 1 - [y + y(1 - y)]$$
$$= (1 - y)^2 \qquad (11.9)$$

Using Eqs. (11.6)–(11.9) for the nth project release, we write

$$m = 1 - (1 - y)^n \qquad (11.10)$$

and

$$d = (1 - y)^n \qquad (11.11)$$

Example 11.2

Assume that after the completion of a project, 15% of the work force is assigned to maintenance each time, and there are seven projects of 1.5 years duration. Calculate the percentage of the total work force that will be assigned to the maintenance of seven projects.

By substituting the given data into Eqs. (11.10) and (11.11), we get

$$m = 1 - (1 - 0.15)^7 = 0.6794$$

and

$$d = (1 - 0.15)^7 = 0.3205$$

It means, approximately 68% of the total work force will be assigned to the maintenance of seven projects.

BELADY–LEHMAN MODEL

This model defines the important relationship among the factors determining maintenance effort as follows:[43]

$$ME = PE + C^{\alpha - \theta} \tag{11.12}$$

where
  ME = total maintenance effort expended for a software system,
  PE = wholly productive efforts that include coding, design, analysis and evaluation, and testing,
  C = an empirical constant and its value depend on the environment,
  α = the complexity caused by the shortcoming of structured design and documentation (the value of α will be high when a system is developed without applying software engineering principles),
  θ = the degree to which the maintenance manpower is familiar with the software under consideration (if the software is maintained without much comprehension, the value of θ will be low).

The model is described in detail in References 22 and 43.

MAINTENANCE COST MODEL

This model can be used to estimate software maintenance cost directly. The software maintenance cost is defined by[44,45]

$$C_{sm} = 3(CMM)(N)/DC \tag{11.13}$$

where

C_{sm} = software maintenance cost,

N = number of instructions to be changed per month,

CMM = cost per man-month,

DC = difficulty constant; its specified values for easy, medium, and hard programs are 500, 250, and 100, respectively.

SOFTWARE MAINTENANCE MANUAL AND STANDARDS ON SOFTWARE MAINTENANCE

A software maintenance manual may be described as a programmer's technical reference guide used as a tool for implementing changes to software. The main objective of this document is to provide program maintenance professionals with general and specific information on the system application software and configuration. Some areas that must be addressed in the manual include testing standards and procedures, maintenance tools, source code standards, change control process, security, and system manual update. A detailed outline of a software maintenance manual is given in Reference 46.

Two important standards on software maintenance are IEEE-STD-1219-1993, *IEEE Standard for Software Maintenance*[47] and International Organization for Standardization/International Electrotechnical Commission Standard ISO/IEC 14764, *Information Technology: Software Maintenance.*[48] The former standard is briefly described below.[2]

The IEEE standard describes in detail the process for managing and executing activities associated with software maintenance. The basis for the standard is a seven-stage activity model of software maintenance: identification of problem, analysis, design, implementation, system testing, acceptance testing, and delivery. There are five attributes associated with each of these seven activities or stages: activity definition, control, input life cycle products, output life cycle products, and metrics.

PROBLEMS

1. Write an essay on software maintenance facts and figures.
2. List six types of software maintenance requests.
3. Describe the following types of software maintenance:
   - Perfective maintenance
   - Adaptive maintenance
   - Corrective maintenance
4. Discuss the problems associated with software maintenance.
5. Define the following terms:
   - Software maintenance
   - Software maintainability
6. What is the difference between the Fog index and the source code readability index?

7. Discuss software configuration management.
8. Describe the following tools directly or indirectly associated with software maintenance:
   - File comparator
   - Text editor
   - Compiler
9. List at least ten factors that can affect the software maintenance cost.
10. What is a software maintenance manual? What are the important areas that must be addressed in a software maintenance manual?

REFERENCES

1. Keene, S.J., Software reliability concepts, Annual Reliability and Maintainability Symposium Tutorial Notes, 1992, 1–21.
2. Bennett, K.H., Software maintenance: a tutorial, in *Software Engineering*, Dorfman, M. and Thayer, R.H., eds., IEEE Computer Society Press, Los Alamitos, California, 1997, 289–303.
3. Boehm, B.W., *Software Engineering Economics*, Prentice-Hall, Englewood Cliffs, New Jersey, 1981.
4. Stevenson, C., *Software Engineering Productivity*, Chapman and Hall, London, 1995.
5. IEEE-Std-610.12-1990, *IEEE Standard Glossary of Software Engineering Terminology*, Institute of Electrical and Electronic Engineers, New York, 1991.
6. Omdahl, T.P., ed., *Reliability, Availability, and Maintainability (RAM) Dictionary*, ASQC Quality Press, Milwaukee, Wisconsin, 1988.
7. Martin, J., *Fourth-Generation Languages*, Vol. 1, Prentice-Hall, Englewood Cliffs, New Jersey, 1985.
8. Foster, J., *Cost Factors in Software Maintenance*, Ph.D. Dissertation, Department of Computer Science, University of Durham, U.K., 1993.
9. Coleman, D., Using metrics to evaluate software system maintainability, *Computer*, 27: 8, 1994, 44–49.
10. Coggins, J.M., *Team Software Engineering and Project Management*, Department of Computer Science, University of North Carolina, Chapel Hill, North Carolina, 1994.
11. Stacey, D., Software Engineering, Course 27-320 Lecture Notes, Department of Computer Science, University of Guelph, Guelph, Ontario, Canada, 1999.
12. Horowitz, B.M., *Strategic Buying for the Future*, Libey Publishing, Washington, D.C., 1993.
13. Fairley, R.E., *Software Engineering Concepts*, McGraw-Hill, New York, 1985.
14. Charette, R.N., *Software Engineering Environments*, Intertext Publications, New York, 1986.
15. Lientz, B.P. and Swanson, E.B., *Software Maintenance Management: A Study of the Maintenance of Computer Application Software in 487 Data Processing Organizations*, Addison-Wesley, Reading, Massachusetts, 1980.
16. Van Tassel, D., *Program, Style, Design, Efficiency, Debugging, and Testing*, Prentice-Hall, Englewood Cliffs, New Jersey, 1978.
17. De Marco, T., *Controlling Software Projects*, Yourdon Press, Englewood Cliffs, New Jersey, 1982.
18. Boeing Company, Software Cost Measuring and Reporting, ASD Document No. D180-22813-1, United States Air Force, Washington, D.C., January 1979.

19. Boehm, B.W., *Software Engineering Economics,* Prentice-Hall, Englewood Cliffs, New Jersey, 1981.
20. Glass, R.L., *Software Reliability Guidebook,* Prentice-Hall, Englewood Cliffs, New Jersey, 1979.
21. Schatz, W., Fed facts, *Datamation,* 15, August 1986, 72–73.
22. Pfleeger, S.L., *Software Engineering: Theory and Practice,* Prentice-Hall, Upper Saddle River, New Jersey, 1998.
23. Lientz, B.P. and Swanson, E.B., Problems in application software maintenance, *Communications of the ACM,* 24: 11, 1981, 763–769.
24. *Pricing Handbook,* Federal Aviation Administration (FAA), Washington, D.C., 1999.
25. Parikh, G. and Zvegintzov, N., *Tutorial on Software Maintenance,* IEEE Computer Society Press, Los Alamitos, California, 1993.
26. McCabe, T., A software complexity measure, *IEEE Transac. Software Engineering,* 2: 4, 1976, 308–320.
27. Shooman, M.L., *Software Engineering,* McGraw-Hill, New York, 1983.
28. Fries, R.C., *Reliable Design of Medical Devices,* Marcel Dekker, New York, 1997.
29. Gunning, R., *The Technique of Clear Writing,* McGraw-Hill, New York, 1968.
30. Dhillon, B.S., *Engineering Design: A Modem Approach,* Richard D. Irwin, Chicago, Illinois, 1996.
31. De Yong, G.E. and Kampen, G.R., Program factors as predictors of program readability, in *Proceedings of the IEEE Computer Software and Applications Conference,* 1979, 668–673.
32. Holbrook, H.B. and Thebaut, S.M., A survey of software maintenance tools that enhance program understanding, Report No. SERC-TR-9-F, Software Engineering Research Center, Department of Computer Science, Purdue University, West Lafayette, Indiana, September 1987.
33. Cashman, P.M. and Holt, A.W., A communication-oriented approach to structuring the software maintenance environment, *ACM SIGSOFT Software Engineering Notes,* 5: 1, January 1980.
34. Pfleeger, S.L. and Bohner, S., A framework for maintenance metrics, in *Proceedings of the IEEE Conference on Software Maintenance,* 1990, 225–230.
35. Schneider, G.R.E., Structured software maintenance, in *Proceedings of the AFIPS National Computer Conference,* 1983, 137–144.
36. Arthur, L.J., *Software Evolution: The Software Maintenance Challenge,* John Wiley & Sons, New York, 1988.
37. Gilb, T., *Principles of Software Engineering Management,* Addison-Wesley, Wokingham, Berkshire, U.K., 1988.
38. Hall, R.P., Seven ways to cut software maintenance costs, *Datamation,* July 15, 1987, 81–84.
39. Lindhorst, W.M., Scheduled maintenance of applications software, *Datamation,* May 1973, 64–67.
40. Parikh, G., Three Ts key to maintenance programming, *Computing S.A.,* April 22, 1981, 19.
41. Yourdon, E., Structured maintenance, in *Techniques of Program and System Maintenance,* G. Parikh, ed., Ethnotech, Lincoln, Nebraska, 1980, 211–213.
42. Mills, H.D., Software development, in *Proceedings of the IEEE Second International Conference on Software Engineering,* Vol. II, 1976, 79–83.
43. Belady, L. and Lehman, M.M., An introduction to growth dynamics, in *Statistical Computer Performance Evaluation,* W. Freiberger, ed., Academic Press, New York, 1972.

44. Dhillon, B.S., *Life Cycle Costing,* Grodon and Breach Science Publishers, New York, 1989.

45. Sheldon, M.R., *Life Cycle Costing: A Better Method of Government Procurement,* Westview Press, Boulder, Colorado, 1979.

46. EPA Directive 2181, *Operation and Maintenance Manual,* Environmental Protection Agency, Washington, D.C.

47. IEEE-Std-1219-1993, *IEEE Standard for Software Maintenance,* Institute of Electrical and Electronic Engineers, New York, 1994.

48. ISO/IEC 14764, *Information Technology: Software Maintenance,* International Organization for Standardization (ISO) International Electrotechnical Commission, Geneva, Switzerland, 1999.

12 Reliability

INTRODUCTION

Reliability is the probability that an item will perform its stated mission satisfactorily for the given time period when used under the specified conditions. It is an important factor in equipment maintenance because lower equipment reliability means higher need for maintenance.

The history of reliability as a discipline may be traced to the 1930s when probability concepts were applied to electric power generation-related problems.[1-4] During World War II, Germans used basic reliability concepts to improve reliability of their V1 and V2 missiles. In 1950, the U.S. Department of Defense established an ad hoc committee on reliability, and in 1952 it was transformed to a permanent group and became known as Advisory Group on the Reliability of Electronic Equipment (AGREE).[5] A report prepared by AGREE was released in 1957.[6]

In 1954, a National Symposium on Reliability and Quality Control was held for the first time in the United States. In 1956, the first commercially available book, entitled *Reliability Factors for Ground Electronic Equipment,* was published.[7] The following year, the U.S. Air Force (USAF) released the first military reliability specification, entitled *Reliability Assurance Program for Electronic Equipment.*[8] In 1962, a graduate degree program in reliability engineering was started by the Air Force Institute of Technology, Dayton, Ohio.

Today, there are many publications available on the discipline of reliability, and each year many conferences are held around the world that deal directly or indirectly with the field. In addition, many academic institutions offer programs in reliability engineering. A comprehensive list of publications on reliability and related areas is given in References 9 and 10.

ROOT CAUSE OF EQUIPMENT RELIABILITY PROBLEMS AND BATHTUB HAZARD RATE CONCEPT

The basic requirement of plant performance is equipment reliability because factors such as product quality, profitability, and production capacity hinge on this crucial factor alone. Over the years various studies have been conducted to determine the root cause of poor equipment reliability. One study based on data collected over a 30-year period categorized the root cause of equipment reliability problems into the following six groups:[11]

1. Sales and marketing: 28%
2. Production scheduling: 20%
3. Maintenance: 17%

TABLE 12.1
Causes of Failure During the Burn-In Period

Failure Cause

Poor quality control
Inadequate materials
Marginal parts
Incorrect use procedures
Poor test specifications
Over-stressed parts
Incorrect installation or setup
Poor manufacturing processes or tooling
Incomplete final test
Wrong handling or packaging
Poor technical representative training
Power surges

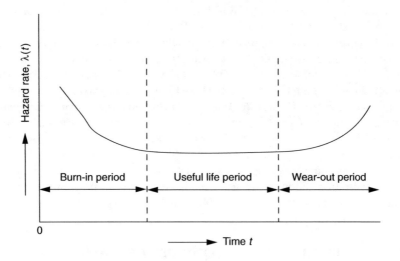

FIGURE 12.1 Bathtub hazard rate curve.

4. Production practices: 17%
5. Purchasing: 10%
6. Plant engineering: 8%

In reliability analysis of engineering systems it is often assumed that the hazard or time-dependent failure rate of items follows the shape of a bathtub as shown in Fig. 12.1.

The curve shown in Fig. 12.1 has three distinct regions: burn-in period, useful life period, and wear-out period. The burn-in period is also known as "infant mortality period," "break-in period," or "debugging period." During this time frame the hazard rate decreases and the failures occur due to causes such as presented in Table 12.1.[12]

In the useful life period the hazard rate is constant and the failures occur randomly or unpredictably. Some of the causes of failures in this region include insufficient design margins, incorrect use environments, undetectable defects, human error and abuse, and unavoidable failures (i.e., ones that cannot be avoided by even the most effective preventive maintenance practices). The wear-out period begins when the item passes its useful life period. During the wear-out period the hazard rate increases. Some causes for the occurrence of wear-out region failures are: wear due to aging, inadequate or improper preventive maintenance, limited-life components, wear due to friction, misalignments, corrosion and creep, and incorrect overhaul practices. Wearout period failures can be reduced significantly by executing effective replacement and preventive maintenance policies and procedures.

RELIABILITY MEASURES

This section presents formulas to obtain item reliability hazard rate and mean time to failure.

RELIABILITY FUNCTION

The reliability of an item can be obtained by using any of the following three equations:[13-14]

$$R(t) = 1 - F(t) = 1 - \int_0^t f(t)dt \tag{12.1}$$

where
$R(t)$ = reliability at time t,
$F(t)$ = cumulative distribution function,
$f(t)$ = failure density function.

$$R(t) = \int_t^\infty f(t)dt \tag{12.2}$$

$$R(t) = \exp\left[-\int_0^t \lambda(t)dt\right] \tag{12.3}$$

where
$\lambda(t)$ = hazard rate or time-dependent failure rate.

Example 12.1

An electric motor times to failure are described by the following probability density function:

$$f(t) = \lambda e^{-\lambda t} \tag{12.4}$$

where
  t = time,
  λ = motor failure rate.

Obtain an expression for motor reliability by using Eqs. (12.1) and (12.2). Comment on the result.

Inserting Eq. (12.4) into Eq. (12.1), we get

$$R(t) = 1 - \int_0^t \lambda e^{-\lambda t} \, dt = e^{-\lambda t} \tag{12.5}$$

Similarly, by substituting Eq. (12.4) into Eq. (12.2) yields

$$R(t) = \int_t^\infty \lambda e^{-\lambda t} \, dt = e^{-\lambda t} \tag{12.6}$$

Eqs. (12.5) and (12.6) are identical, and prove that both Eqs. (12.1) and (12.2) give the same result.

HAZARD RATE

This is defined by

$$\lambda(t) = \frac{f(t)}{R(t)} = -\frac{1}{R(t)} \cdot \frac{dR(t)}{dt} \tag{12.7}$$

Example 12.2

Obtain a hazard rate expression by using Eqs. (12.4) and (12.5) in Eq. (12.7). Use the resulting hazard rate expression in Eq. (12.3) to get an expression for reliability. Comment on the result.

Using Eqs. (12.4) and (12.5) in Eq. (12.7) yields

$$\lambda(t) = \frac{\lambda e^{-\lambda t}}{e^{-\lambda t}} = \lambda \tag{12.8}$$

Substituting Eq. (12.8) into Eq. (12.3), we get

$$R(t) = \exp\left[-\int_0^t \lambda \, dt\right] = e^{-\lambda t} \tag{12.9}$$

The above equation is identical to Eqs. (12.5) and (12.6). This proves that Eqs. (12.1)–(12.3) yield identical results.

MEAN TIME TO FAILURE (MTTF)

This is defined by

$$\text{MTTF} = \int_0^\infty R(t)\,dt \qquad (12.10)$$

The following two expressions also yield the identical result:

$$\text{MTTF} = \int_0^\infty t f(t)\,dt \qquad (12.11)$$

$$\text{MTTF} = \lim_{s \to 0} R(s) \qquad (12.12)$$

where

s = Laplace transform variable,

$R(s)$ = Laplace transform of the reliability function, $R(t)$.

Example 12.3

Assume that the reliability of a mechanical device is defined by

$$R(t) = e^{-\lambda t} \qquad (12.13)$$

where

$\lambda = 0.0004$ failures per hour.

Calculate the device MTTF by using Eqs. (12.10) and (12.12). Comment on the end result.

Using Eq. (12.13) in Eq. (12.10) yields:

$$\text{MTTF} = \int_0^\infty e^{-\lambda t}\,dt$$

$$= \frac{1}{\lambda}$$

$$= \frac{1}{0.0004}$$

$$= 2500 \text{ h}$$

By taking Laplace transform of Eq. (12.13) we get

$$R(s) = \frac{1}{s + \lambda} \qquad (12.14)$$

TABLE 12.2
Formulas for Obtaining Item Reliability, Hazard Rate, and MTTF

No	Reliability	Hazard Rate	MTTF
1	$R(t) = 1 - \int_0^t f(t)\,dt$	$\lambda(t) = \dfrac{f(t)}{R(t)}$	$\int_0^\infty t f(t)\,dt$
2	$R(t) = \int_t^\infty f(t)\,dt$	$\lambda(t) = -\dfrac{1}{R(t)}\dfrac{dR(t)}{dt}$	$\int_0^\infty R(t)\,dt$
3	$R(t) = \exp\left[-\int_0^t \lambda(t)\,dt\right]$	$\lambda(t) = \dfrac{f(t)}{\int_t^\infty f(t)\,dt}$	$\lim_{s\to 0} R(s)$

By inserting Eq. (12.14) into Eq. (12.12), we obtain

$$MTTF = \lim_{s\to 0}\left(\frac{1}{s+\lambda}\right)$$

$$= \frac{1}{\lambda}$$

$$= \frac{1}{0.0004}$$

$$= 2500 \text{ h}$$

Eqs. (12.10) and (12.12) give identical results for the mechanical device mean time to failure, i.e., 2500 h. Table 12.2 presents formulas to obtain item reliability, hazard rate, and MTTF.

RELIABILITY NETWORKS

This section is concerned with the reliability evaluation of most standard networks occurring in engineering systems. The networks covered in this section are series, parallel, and standby.

SERIES NETWORK

In this case n number of units forms a series system, as shown in Fig. 12.2. If any one of the units fails, the system fails. All system units must work normally for successful operation of the system.

A typical example of a series system is four wheels of a car. If any one of the tires punctures, the car for practical purposes cannot be driven. Thus, these four tires form a series system. For independent and nonidentical units, the series system, shown in Fig. 12.2, reliability is

$$R_S = R_1 R_2 R_3 \ldots R_n \tag{12.15}$$

FIGURE 12.2 The block diagram of an n-unit series system.

where
   R_S = series system reliability,
   n = number of units,
   R_i = reliability of unit or block i shown in Fig. 12.2, for $i = 1, 2, 3, ..., n$.

For exponentially distributed times to failure of unit i, the unit reliability from Eq. (12.5) can be written as

$$R_i(t) = e^{-\lambda_i t} \tag{12.16}$$

where
   $R_i(t)$ = reliability of unit i at time t, for $i = 1, 2, 3, ..., n$,
   λ_i = constant failure rate of unit i, for $i = 1, 2, 3, ..., n$.

   By substituting Eq. (12.16) into Eq. (12.15), we get

$$R_S(t) = e^{-\sum_{i=1}^{n} \lambda_i t} \tag{12.17}$$

where
   $R_S(t)$ = series system reliability at time t.

   Using Eq. (12.17) in Eq. (12.10) yields

$$MTTF_S = \int_0^\infty e^{-\sum_{i=1}^{n} \lambda_i t} dt$$

$$= \frac{1}{\sum_{i=1}^{n} \lambda_i} \tag{12.18}$$

where
   $MTTF_S$ = series system mean time to failure.

   Inserting Eq. (12.17) into Eq. (12.7), we obtain

$$\lambda_S(t) = \left(\frac{1}{e^{-\sum_{i=1}^{n} \lambda_i t}} \right) \frac{d e^{-\sum_{i=1}^{n} \lambda_i t}}{dt} = \sum_{i=1}^{n} \lambda_i \tag{12.19}$$

where
$\lambda_S(t)$ = series system hazard (failure) rate.

Example 12.4

Assume that the constant failure rates of tires 1, 2, 3, and 4 of a car are $\lambda_1 = 0.00001$ failures per hour, $\lambda_2 = 0.00002$ failures per hour, $\lambda_3 = 0.00003$ failures per hour, and $\lambda_4 = 0.00004$ failures per hour, respectively. For practical purposes, the car cannot be driven when any one of the tires punctures. Calculate the total tire system failure rate and mean time to failure of the car with respect to tires.

Substituting the given data into Eq. (12.19) yields

$$\lambda_S = 0.00001 + 0.00002 + 0.00003 + 0.00004$$
$$= 0.0001 \text{ failures per hour}$$

Using the above result in Eq. (12.18) we get

$$\text{MTTF}_S = \frac{1}{0.0001} = 10,000 \text{ h}$$

The total tire system failure rate and mean time to failure of the car with respect to tires are 0.0001 failures per hour and 10,000 h, respectively.

PARALLEL NETWORK

In this case n number of simultaneously operating units form a parallel system, as shown in Fig. 12.3. Each block in the figure denotes a unit. At least one of the units must work normally for system success.

For independent units, the parallel system shown in Fig. 12.3 reliability is given by

$$R_{ps} = 1 - (1 - R_1)(1 - R_2)(1 - R_3)...(1 - R_n) \qquad (12.20)$$

where
R_{ps} = parallel system reliability,
n = total number of units,
R_i = reliability of unit i, for $i = 1, 2, 3,...,n$.

For exponentially distributed times to failure of unit i, substitute Eq. (12.16) into Eq. (12.20) to get

$$R_{ps}(t) = 1 - \prod_{i=1}^{n} (1 - e^{-\lambda_i t}) \qquad (12.21)$$

where
$R_{ps}(t)$ = parallel system reliability at time t.

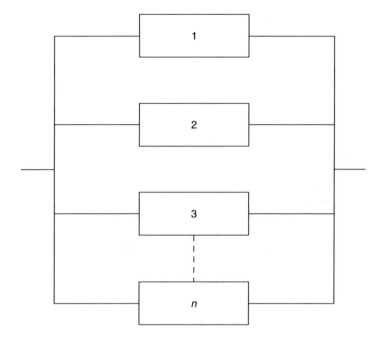

FIGURE 12.3 An n-unit parallel system.

For identical units (i.e., $\lambda_i = \lambda$), Eq. (12.21) simplifies

$$R_{ps}(t) = 1 - (1 - e^{-\lambda t})^n \qquad (12.22)$$

where λ is the unit failure rate.

By substituting Eq. (12.22) into Eq. (12.10) we get

$$\text{MTTF}_{ps} = \int_0^{\infty} [1 - (1 - e^{-\lambda t})^n] dt = \frac{1}{\lambda} \sum_{i=1}^{n} \frac{1}{i} \qquad (12.23)$$

where
MTTF_{ps} = identical unit parallel system mean time to failure.

Example 12.5

An aircraft has two independent and active engines. At least one engine must operate normally for the aircraft to fly. Engines 1 and 2 reliabilities are 0.99 and 0.97, respectively. Calculate the probability of the aircraft flying successfully with respect to engines.

Substituting the given data into Eq. (12.20) yields

$$R_{ps} = 1 - (1 - 0.99)(1 - 0.97) = 0.9997$$

There is 99.97% chance that the aircraft will fly successfully with respect to engines.

Example 12.6

A system is composed of two independent and active units, and at least one unit must work normally for the system success. The constant failure rates of units 1 and 2 are $\lambda_1 = 0.004$ failures per hour and $\lambda_2 = 0.006$ failures per hour, respectively. Calculate the system mean time to failure.

By substituting the given data into Eq. (12.21), we obtain

$$R_{ps} = 1 - (1 - e^{-0.004t})(1 - e^{-0.006t})$$
$$= e^{-0.004t} + e^{-0.006t} + e^{-(0.004 + 0.006)t} \tag{12.24}$$

Substituting Eq. (12.24) into Eq. (12.10), we get

$$MTTF_{ps} = \int_0^\infty [e^{-0.004t} + e^{-0.006t} - e^{-(0.004 + 0.006)t}] dt$$
$$= \frac{1}{0.004} + \frac{1}{0.006} - \frac{1}{0.004 + 0.006}$$
$$= 316.67 \text{ h}$$

System mean time to failure is 316.67 h. More specifically, expect a failure after every 316.67 h of operation.

STANDBY SYSTEM

In this case one unit is operating and k units are in standby mode. As soon as the operating unit fails, it is immediately replaced with one of the standby units. The system has a total of $(k + 1)$ units. Figure 12.4 shows a block diagram of a standby system with $(k + 1)$ units. Each block in the figure denotes a unit.

The standby system reliability is given by[15,16]

$$R_{sbs}(t) = \sum_{i=0}^{k} \left\{ \left[\int_0^t \lambda(t)dt \right]^i e^{-\int_0^t \lambda(t)dt} \right\} / i! \tag{12.25}$$

where
   R_{sbs} = standby system reliability at time t.

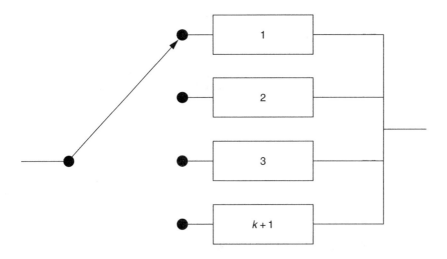

FIGURE 12.4   An $(k + 1)$-unit standby system.

Equation (12.25) is subject to the following assumptions:

- The switching mechanism is perfect.
- All system units are independent and identical.
- The standby units remain as good as new in their standby mode.
- The unit failure rate is nonconstant. Times to failure can be represented
  by any statistical distribution (e.g., Weibull, gamma, or exponential).

For exponentially distributed unit times to failure, set $\lambda(t) = \lambda$ in Eq. (12.25) to get

$$R_{sbs}(t) = \sum_{i=1}^{k}[(\lambda t)^i e^{-\lambda t}]/i! \qquad (12.26)$$

Using Eq. (12.26) in Eq. (12.10) yields

$$MTTF_{sbs} = \int_0^\infty \left[\left\{\sum_{i=0}^{k}(\lambda t)^i e^{\lambda t}\right\}\Big/ i!\right]dt$$

$$= (k + 1)/\lambda \qquad (12.27)$$

where
   $MTTF_{sbs}$ = standby system mean time to failure.

Example 12.7

A system is composed of two independent and identical units — one working, one
on standby. The standby switching mechanism is perfect and the unit failure rate
is 0.0005 failures per hour. Calculate the system mean time to failure and reliability

TABLE 12.3
MTTF Formulas for Selected Reliability Networks

Reliability Network/System	MTTF with Nonidentical Units	MTTF with Identical Units
Series	$\dfrac{1}{\sum\limits_{i=1}^{n} \lambda_i}$	$\dfrac{1}{n\lambda}$
Parallel	–	$\dfrac{1}{\lambda}\sum\limits_{i=1}^{n}\dfrac{1}{i}$
Standby	–	$\dfrac{k+1}{\lambda}$

for a 100-h mission. Assume the standby unit remains as good as new in its standby mode.

Inserting the given data values into Eq. (12.26), we obtain

$$R_{\text{sbs}}(t) = \sum_{i=0}^{1}\{(\lambda t)^i e^{-\lambda t}\}/i!$$

$$= e^{-(0.0005 \times 100)}[1 + (0.0005 \times 100)]$$

$$= 0.9988$$

Using the specified data in Eq. (12.27) yields

$$\mathrm{MTTF}_{\text{sbs}} = \frac{1+1}{0.0005} = 4000\,\text{h}$$

The system reliability and mean time to failure are 0.9988 and 4000 h, respectively.

Table 12.3 presents mean time to failure formulas for some independent unit standard reliability networks.

RELIABILITY ANALYSIS METHODS

Over the years many reliability analysis methods have been developed. This section presents three such commonly used methods in the industrial sector: Markov, fault tree analysis (FTA), and failure modes and effect analysis.

MARKOV METHOD

The Markov method is a powerful reliability analysis method named for a Russian mathematician (1856–1922). The Markov method is a useful tool to model systems with dependent failure and repair modes and constant failure and repair rates. It can also handle some systems having time-dependent failure and repair rates.[14] The Markov method is based on the following assumptions:[14,17]

- The transitional probability from one system state to another in the time Δt is given by $\lambda \Delta t$. The parameter λ is a constant and its dimensions are occurrences per unit time. In reliability work, this constant could be a failure or repair rate.
- The occurrences are independent of all other occurrences.
- The transition probability of two or more occurrences in time interval Δt from one system state to another is negligible (e.g., $(\lambda \Delta t)(\lambda \Delta t) \to 0$).

The application of Markov method is demonstrated through the following two examples.

Example 12.8

A compressor system times to failure are exponentially distributed. The compressor system failure rate, λ_c, is constant. The compressor system transition diagram is shown in Fig. 12.5. Develop expressions for the system reliability and failure probability using the Markov approach. The numerals in boxes denote system state.

With the aid of Markov method, we write the following difference equations for the diagram shown in Fig. 12.5.

$$P_0(t + \Delta t) = P_0(t)(1 - \lambda_c \Delta t) \tag{12.28}$$

$$P_1(t + \Delta t) = P_1(t) + P_0(t)(\lambda_c \Delta t) \tag{12.29}$$

where
$\lambda_c \Delta t$ = probability of compressor system failure in time Δt,
$P_i(t)$ = probability that the compressor system is in state i at time t, for $i = 0, 1$,
$P_i(t + \Delta t)$ = probability of the compressor system being in state i at time $t + \Delta t$, for $i = 0, 1$,
$(1 - \lambda_c \Delta t)$ = probability of no failure in time Δt when the compressor system is in state 0 at time t.

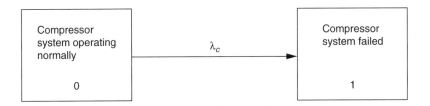

FIGURE 12.5 Compressor system state space diagram.

Rearranging Eqs. (12.28) and (12.29), we get

$$\lim_{\Delta t \to 0} \frac{P_0(t + \Delta t) - P_0(t)}{\Delta t} = \frac{dP_0(t)}{dt} = -\lambda_c P_0(t) \qquad (12.30)$$

$$\lim_{\Delta t \to 0} \frac{P_1(t + \Delta t) - P_1(t)}{\Delta t} = \frac{dP_1(t)}{dt} = \lambda_c P_0(t) \qquad (12.31)$$

At time $t = 0$, $P_0(0) = 1$ and $P_1(0) = 0$.

Solving Eqs. (12.30) and (12.31) with the Laplace transform approach, we get

$$P_0(s) = \frac{1}{s + \lambda_c} \qquad (12.32)$$

$$P_1(s) = \frac{\lambda_c}{s(s + \lambda_c)} \qquad (12.33)$$

Taking the inverse Laplace transforms of Eqs. (12.32) and (12.33), we obtain

$$R_c(t) = P_0(t) = e^{-\lambda_c t} \qquad (12.34)$$

$$F_c(t) = P_1(t) = 1 - e^{-\lambda_c t} \qquad (12.35)$$

where
   $R_c(t)$ = compressor system reliability at time t,
   $F_c(t)$ = compressor system failure probability at time t.

Example 12.9

Assume a parallel system is composed of two independent and identical units. As soon as a unit fails, it is immediately repaired at a rate, μ. The total system can also fail due to the occurrence of a common-cause failure.[18] Occurrence of a common-cause failure leads to the simultaneous failure of both units. The system state space diagram is shown in Fig. 12.6. The numerals in boxes denote system state. Obtain an expression for the system mean time to failure.
   The system is subject to the following assumptions:

- The unit failure rate is constant.
- The system common-cause failure rate is constant.
- Both the units are operating simultaneously.
- A failed unit is repaired.
- The failed system is never repaired.

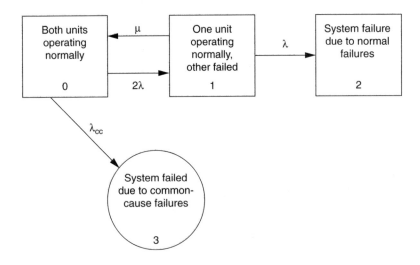

FIGURE 12.6 Transition diagram of a two-unit parallel system with common-cause failures.

The following symbols were used to develop equations for the Fig. 12.6 diagram:

i = the ith system state, for $i = 0$ (both units operating normally), $i = 1$ (one unit operating normally, other failed), $i = 2$ (system failed due to normal failures), $i = 3$ (system failed due to common-cause failures),

$P_i(t)$ = probability that the system is in state i at time t, for $i = 0, 1, 2, 3$,

λ = constant unit failure rate,

λ_{cc} = constant system common-cause failure rate,

$P_i(s)$ = Laplace transform of the probability that the system is in state i, for $i = 0, 1, 2, 3$,

μ = constant unit repair rate.

With the aid of the Markov method, we write the following equations for Fig. 12.6 diagram:

$$\frac{dP_0(t)}{dt} + (2\lambda + \lambda_{cc})P_0(t) = \mu P_1(t) \tag{12.36}$$

$$\frac{dP_1(t)}{dt} + (\lambda + \mu)P_1(t) = 2\lambda P_0(t) \tag{12.37}$$

$$\frac{dP_2(t)}{dt} = \lambda P_1(t) \tag{12.38}$$

$$\frac{dP_3(t)}{dt} = \lambda_{cc}P_0(t) \tag{12.39}$$

At time $t = 0$, $P_0(0) = 1$ and $P_1(0) = P_2(0) = P_3(0) = 0$.

Solving Eqs. (12.36)–(12.39) with Laplace transforms, we get

$$P_0(s) = \frac{s + \lambda + \mu}{A} \tag{12.40}$$

where

$$A \equiv (s + 2\lambda + \lambda_{cc})(s + \lambda + \mu) - 2\lambda\mu$$

$$P_1(s) = 2\lambda / A \tag{12.41}$$

$$P_2(s) = 2\lambda^2 / sA \tag{12.42}$$

$$P_3(s) = \lambda_{cc}(s + \lambda + \mu) / sA \tag{12.43}$$

The Laplace transform of system reliability is given by

$$R_{ps}(s) = P_0(s) + P_1(s) = [(s + \lambda + \mu) + 2\lambda] / A \tag{12.44}$$

where $R_{ps}(s)$ is the Laplace transform of the two-unit parallel system reliability.

Substituting Eq. (12.44) into Eq. (12.12), we get

$$\text{MTTF}_{ps} = \lim_{s \to 0} [(s + \lambda + \mu) + 2\lambda] / A$$

$$= \frac{3\lambda + \mu}{2\lambda^2 + \lambda\lambda_{cc} + \lambda_{cc}\mu} \tag{12.45}$$

where
MTTF_{ps} = two-unit parrallel system mean time to failure.

Fault Tree Analysis

Fault tree analysis (FTA) is one of the most widely used methods in the industrial sector to perform reliability analysis of complex engineering systems. A fault tree is a logical representation of the relationship of primary/basic events that lead to a given undesirable event (i.e., top event). It is depicted using a tree structure with logic gates such as OR and AND.

FTA was developed in the early 1960s at Bell Labs to perform reliability analysis of the Minuteman Launch Control System.[19,20] A comprehensive list of publications on FTA is given in Reference 9.

FTA begins with identification of an undesirable event called the top event of a given system. Fault events which could make the top event occur are generated and connected by logic gates such as OR and AND.

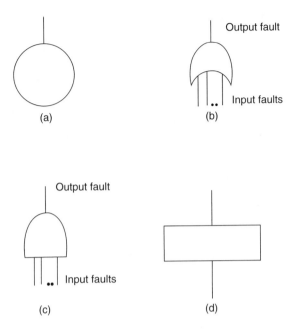

FIGURE 12.7 Basic fault tree symbols: (a) basic fault event, (b) OR gate, (c) AND gate, (d) resultant fault event.

The OR gate provides a TRUE (failure) output if one or more of its input faults are present. In contrast, the AND gate provides a TRUE (failure) output if all of its input faults are present. Symbols for both OR and AND gates are shown in Fig. 12.7.

The fault tree construction proceeds by generation of fault events successively until the fault events need not be developed further. These fault events are known as basic fault events and the fault tree itself is the logic structure relating the top fault event to the basic fault events.

Four basic symbols used in fault tree construction are shown in Fig. 12.7. The meanings of both OR and AND gate symbols were discussed earlier. Circle and rectangle symbols denote a basic fault event and the resultant fault event which occur from the combination of fault events through the input of a gate, respectively.

The development or construction of a fault tree is a top-down process (i.e., starting from the top event moving downward). It consists of successively asking the question, "How could this event occur?" The following basic steps are involved in performing FTA:[21]

- Define factors such as system assumptions, and what constitutes a failure.
- Develop system a block diagram showing items such as interfaces, inputs, and outputs.
- Identify undesirable or top fault event.
- Using fault tree symbols, highlight all causes that can make the top event occur.
- Construct the fault tree to the lowest level required.

- Analyze the fault tree as per the requirements.
- Identify necessary corrective measures.
- Document and followup on highlighted corrective measures.

The following example demonstrate the development of a fault tree.

Example 12.10

A room has two light bulbs and one switch. Develop a fault tree for the top event — room not lit. Assume the following:

- The room is windowless.
- The switch can only fail to close.
- The room will only become dark if there is no electricity, both light bulbs burn out, or the switch fails to close.

A fault tree for the example is shown in Fig. 12.8. Each event in the figure is labeled E_1, E_2, E_3, E_4,…, E_8.

Probability Evaluation

The probability of OR and AND gate output fault event occurrence can be calculated using Eqs. (12.46) and (12.47) below.

$$P(E_0) = 1 - \prod_{i=1}^{k}\{1 - P(E_i)\} \qquad (12.46)$$

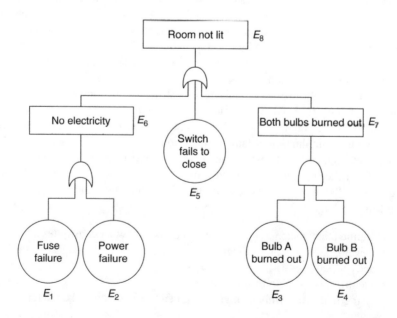

FIGURE 12.8 A fault tree for Example 12.10.

and

$$P(E_a) = \prod_{i=1}^{k} P(E_i) \tag{12.47}$$

where
  $P(E_0)$ = probability of occurrence of OR gate output fault event,
  $P(E_a)$ = probability of occurrence of AND gate output fault event,
  k = total number of input fault events,
  $P(E_i)$ = probability of occurrence of input fault event E_i, for $i = 1, 2, 3,..., k$.

Example 12.11

Assume that in Fig. 12.8 the probabilities of occurrence of events E_1, E_2, E_3, E_4, and E_5 are 0.04, 0.05, 0.06, 0.07, and 0.08, respectively. Calculate the probability of occurrence of the top event — room not lit.

Using Eqs. (12.46) and (12.47) and the given data, we get

$$P(E_6) = 1 - (1 - 0.04)(1 - 0.05) = 0.088$$

and

$$P(E_7) = 0.06 \times 0.07 = 0.0042$$

Using the above calculated values, the given data value, and Eq. (12.46) yields

$$P(E_8) = 1 - (1 - 0.08)(1 - 0.088)(1 - 0.0042) = 0.1645$$

There is 16.45% chance the room is not lit.

FAILURE MODES AND EFFECT ANALYSIS

Failure modes and effect analysis (FMEA) is one of the most widely used methods to evaluate design at the initial stage from the reliability aspect. The technique helps identify requirements for and the effects of design change. This method was developed in the early 1950s to evaluate the design of flight control systems from the reliability aspect.[22,23]

FMEA demands listing potential failure modes of each system/equipment/device/part on paper and its effects on the listed subsystems/systems/parts/etc. The fundamental difference between FMEA and FTA is that the former is failure-oriented and the latter event-oriented.

The basic steps used in performing FMEA are as follows:

1. Define system/equipment/item boundaries and associated detailed requirements.
2. List all system/item components and subsystems.

3. Identify each component, its associated failure modes, and their descriptions.
4. Assign failure rate/probability to each identified component failure mode.
5. List effect or effects of each failure mode on subsystem/plant.
6. Enter remarks for each failure mode.
7. Review each critical failure mode and initiate appropriate measures.

The FMEA method is described in detail in Reference 4.

PROBLEMS

1. Write an essay on the early developments in the reliability field.
2. What are the root causes of equipment reliability problems?
3. Describe the bathtub hazard rate concept.
4. What are the failure causes for the burn-in period of a bathtub hazard rate curve?
5. Write three expressions for obtaining mean time to failure of an item.
6. What is the difference between hazard rate and failure rate?
7. Prove that the mean time to failure of a parallel system is given by

$$\text{MTTF}_p = \frac{1}{\lambda_1} + \frac{1}{\lambda_2} - \frac{1}{\lambda_1 + \lambda_2} \qquad (12.48)$$

where λ_1 is the unit 1 failure rate and λ_2 is the unit 2 failure rate.
State any assumption made in your proof.
8. Prove that a standby system reliability is given by

$$R_{sbs}(t) = \sum_{i=0}^{1} \{(\lambda t)^i e^{-\lambda t}\}/i! \qquad (12.49)$$

where λ is the unit failure rate and t is time.
State all the assumptions made in your proof.
9. Write three assumptions on which the Markov method is based.
10. Describe the following reliability analysis methods:
   • FTA
   • FMEA

REFERENCES

1. Smith, S.A., Service reliability measured by probabilities of outage, *Electrical World,* 103, 1934, 371–374.
2. Lyman, W.J., Fundamental consideration in preparing a master system plan, *Electrical World,* 101, 1933, 778–792.
3. Dhillon, B.S., *Power System Reliability, Safety and Management,* Ann Arbor Science Publishers, Ann Arbor, Michigan, 1983.

4. Dhillon, B.S., *Design Reliability: Fundamentals and Applications,* CRC Press, Boca Raton, Florida, 1999.
5. Cappola, A., Reliability engineering of electronic equipment: a historical perspective, *IEEE Transac. Reliability,* 33, 1984, 29–35.
6. AGREE Report, Advisory Group on Reliability of Electronic Equipment, Reliability of Military Electronic Equipment, Office of the Assistant Secretary of Defense (Research and Engineering), Department of Defense, Washington, D.C., 1957.
7. Henny, K., ed., *Reliability Factors for Ground Electronic Equipment,* McGraw-Hill, New York, 1956.
8. MIL-R-25717 (USAF), *Reliability Assurance Program for Electronic Equipment,* Department of Defense, Washington, D.C., 1957.
9. Dhillon, B.S., *Reliability and Quality Control: Bibliography on General and Specialized Areas,* Beta Publishers, Gloucester, Ontario, Canada, 1992.
10. Dhillon, B.S., *Reliability Engineering Applications: Bibliography on Important Application Areas,* Beta Publishers, Gloucester, Ontario, Canada, 1992.
11. Mobley, K., Equipment reliability: who is responsible?, *Plant Services Magazine,* February 1997, 12–15.
12. Ekings, D., Reliability in production, in *Handbook of Reliability Engineering and Management,* W. Grant Ireson and C.F. Coombs, eds., McGraw-Hill, New York, 1988, 7.1–7.30.
13. Thomason, T., *Introduction to Reliability and Quality,* Machinery Publishing Co., London, 1969.
14. Shooman, M.L., *Probabilistic Reliability: An Engineering Approach,* McGraw-Hill, New York, 1968.
15. Sandler, G.H., *System Reliability Engineering,* Prentice-Hall, Englewood Cliffs, New Jersey, 1963.
16. Dhillon, B.S., *Reliability Engineering in Systems Design and Operation,* Van Nostrand Reinhold Co., New York, 1983.
17. Gnedenko, B.V., Belyaev, Y.K., and Solovyev, A.D., *Mathematical Methods of Reliability Theory,* Academic Press, New York, 1969.
18. Jacobs, I.M., The common-mode failure study discipline, *IEEE Trans. Nuclear Science,* 17, 1970, 594–598.
19. *Fault Tree Handbook,* Report No. NUREG-0492, U.S. Nuclear Regulatory Commission, Washington, D.C., 1981.
20. Dhillon, B.S. and Singh, C., Bibliography of literature on fault trees, *Microelectronics and Reliability,* 17, 1978, 501–503.
21. Grant Ireson, W., Coombs, C.F., and Moss, R.Y., ed., *Handbook of Reliability Engineering and Management,* McGraw-Hill, New York, 1996.
22. Countinho, J.S., Failure effect analysis, *Transac. New York Academy of Sciences,* 26, 1964, 564–584.
23. Arnzen, H.E., Failure Mode and Effect Analysis: A Powerful Engineering Tool for Component and System Optimization, Report No. 347.40.00.00-K4-05 (C5776), GIDEP Operations Center, Corona, California, 1966.

13 Maintainability

INTRODUCTION

Maintainability is a design parameter intended to reduce repair time, as opposed to maintenance, which is the act of repairing or servicing an item or equipment.[1]

The history of maintainability can be traced back to 1901 when the U.S. Army Signal Corps contracted for the development of the Wright brothers' airplane contained a clause that the aircraft should be "simple to operate and maintain."[2] In modern context, the real beginning of maintainability could be considered as the 1950s because of the following two events:[3–5]

- In 1956, a 12-part series of articles appeared in *Machine Design*. The series covered areas such as: designing electronic equipment for maintainability, design of covers and cases, design of maintenance controls, designing for installation, design recommendations for test points, and recommendations for designing maintenance access in electronic equipment.
- In 1957, the Advisory Group on Reliability of Electronic Equipment (AGREE), established by the U.S. Department of Defense, released its report containing many recommendations that served as a basis for the majority of standards on maintainability.

The first commercially available book on maintainability appeared in 1960,[6] and in 1966 three U.S. military documents concerning maintainability were released: MIL-HDBK-472 (*Maintainability Prediction*),[7] MIL-STD-470 (*Maintainability Program Requirements*),[8] and MIL-STD-471 (*Maintainability Demonstration*).[9] Since then, many publications on maintainability have appeared, and a comprehensive list of publications on the subject is given in Reference 10.

MAINTAINABILITY TERMS AND DEFINITIONS, IMPORTANCE, AND OBJECTIVES

Some of the terms and definitions associated with maintainability are as follows:[2,13–16]

- *Maintainability:* The probability that a failed item/equipment will be restored to acceptable working condition.
- *Maintainability engineering:* An application of scientific knowledge and skills to develop equipment/item that is inherently able to be maintained

as measured by favorable maintenance characteristics as well as figures-of-merit.

- *Maintainability model:* A quantified representation of a test/process to perform an analysis of results that determine useful relationships between a group of maintainability parameters.
- *Downtime:* The total time in which the item/equipment is not in a satisfactory operable condition.
- *Serviceability:* The degree of ease/difficulty with which an item/equipment can be restored to its satisfactory operable state.
- *Maintainability function:* A plot of the probability of repair within a time given on the *y*-axis, against maintenance time on the *x*-axis and is useful to predict the probability that repair will be completed in a specified time.

There are many factors responsible for the importance of maintainability. In particular, alarmingly high operating and support costs, due to failures and subsequent maintenance, are among the most pressing problems. These problems were even more apparent in the early days of the maintainability field. For example, in the 1950s, the U.S. Air Force performed a study and found that almost one-third of all Air Force personnel were occupied with maintenance, and the entire maintenance activity accounted for approximately one-third of all Air Force operating costs.[17]

The main objective of maintainability is to maximize equipment and facility availability. The other maintainability objectives include: reduce predicted maintenance time and costs by simplifying maintenance through design, determine labor-hours and other resources needed to perform the projected maintenance, and use maintainability data to determine item availability/unavailability.[5,18]

MAINTAINABILITY MANAGEMENT IN SYSTEM LIFE CYCLE

An efficient and effective design can only be achieved by seriously considering maintainability issues that arise during the system life cycle. This means a maintainability program must incorporate a dialogue between the manufacturer and user throughout the system life cycle. This dialogue concerns the user's maintenance needs and other requirements for the system and the manufacturer's response to these needs and requirements.

The life cycle of a system can be divided into the following four phases:[2]

- Phase I: Concept development
- Phase II: Validation
- Phase III: Production
- Phase IV: Operation

Specific maintainability functions concerning each of these phases are discussed below.

Phase I: Concept Development

In Phase I, high risk areas are identified and system operation needs are translated into a set of operational requirements. The primary maintainability concern during this phase is the determination of system effectiveness needs and criteria, in addition to establishment of the maintenance and logistic support policies and boundaries required to satisfy mission objectives by using operational and mission profiles.

Items such as the following must be accomplished prior to developing system maintainability requirements:

- Details of mission, system operating modes, and so on
- Evaluation of system utilization rates and mission time factors
- Details of the global logistic support objectives and concepts
- Evaluation of the system life cycle duration

Phase II: Validation

During Phase II, operational requirements developed and formulated in the previous phase are refined further with respect to system design requirements. The prime objective of validation is to ensure that full-scale development does not begin until factors such as costs, performance and support objectives, and schedules have been effectively prepared and evaluated.

In this phase, maintainability management specifically deals with tasks such as those listed below:

- Preparing maintainability program and demonstration plans as per contractual requirements
- Determining reliability, maintainability, and system effectiveness-related requirements
- Preparing maintainability policies and procedures for validation and follow-on full-scale engineering effort
- Coordinating and monitoring the entire organization's maintainability effort
- Performing maintainability predictions and allocations
- Participating in trade-off analyses
- Providing assistance to maintenance engineering in the performance of maintenance-related analyses
- Preparing plans for data collection and analysis
- Establishing maintainability incentives and penalties
- Participating in design reviews with respect to maintainability
- Developing maintainability design-related guidelines for use by design engineers with the aid of maintenance engineering analyses

Phase III: Production

In Phase III, the system is manufactured, tested, and delivered, and, in some cases, installed per the technical data package resulting from Phases I and II. Although the maintainability engineering design efforts will largely be completed by this time, the maintainability-related tasks such as those listed below are performed during this phase.

- Monitoring the entire production process
- Examining production test trends with respect to adverse effects on items such as maintainability, maintenance concepts, and provisioning plans
- Examining change proposals with respect to their impact on maintainability
- Assuring the proper correction of discrepancies that can adversely impact maintainability
- Taking part in establishment of controls for process variations, errors, etc., that can undermine system maintainability

PHASE IV: OPERATION

In Phase IV, the system is used, logistically supported, and modified as appropriate. During the operation phase maintenance, overhaul, training, supply, and material readiness requirements and characteristics of the system become clear. Although there are no particular maintainability requirements at this time, the phase is probably the most crucial because the actual cost-effectiveness and logistic support of the system are demonstrated. In addition, maintainability-related data can be obtained from the real life experience for future use.

MAINTAINABILITY DESIGN CHARACTERISTICS AND SPECIFIC CONSIDERATIONS

There are many maintainability-related system/item characteristics that must be emphasized during design. Some of these are: modular design, interchangeability, displays, human factors, safety, test points, standardization, controls, illumination, weight, lubrication, accessibility, installation, training needs, adjustments and calibration, tools, labeling and coding, test equipment, manuals, work environment, covers and doors, size and shape, failure indication (location), connectors, and test hookups and adapters. The most commonly cited/mentioned maintainability-related characteristics by professionals involved with maintainability include: displays, controls, doors, covers, labeling and coding, accessibility, test points, checklists, mounting and fasteners, handles, connectors, test equipment, charts, aids, and manuals.[2] Some of these factors are discussed below.

ACCESSIBILITY

This may be described as the relative ease with which an item can be reached for replacement, service, or repair. Inaccessibility is a frequent cause of ineffective maintenance, thus an important maintainability problem. Many factors can affect accessibility. Some of them are as follows:[19]

- Location of item and its associated environment
- Frequency of entering access opening

- Distance to be reached to access the part of component
- Type of maintenance tasks to be performed through the access opening
- Visual needs of personnel performing the tasks
- Types of tools and accessories required to conduct the specified tasks
- Work clearances appropriate for carrying out the specified tasks
- Degree of danger involved in using access opening
- Mounting or packaging of items/parts behind the access opening
- Required times for performing the specified tasks
- Type of clothes worn by the involved personnel

Some guidelines for the design of access openings are as follows:[19]

- Design access openings for maximum convenience in performing the required maintenance tasks.
- Design access openings so they are a safe distance from hazardous moving parts or high voltage points.
- Ensure that access openings occupy the same face as associated features such as displays, controls, and test points.
- Ensure that the location of access openings allows direct access to the parts or components that will subsequently require some kind of maintenance.
- Ensure that access openings will be accessible effectively under normal installation of the equipment or system.
- Ensure that the lower edge of a restricted access opening is no less than 24 in. or its upper edge no more than 60 in. from the work platform or floor.
- Ensure that heavy parts/units can easily be pulled out rather than lifted out.
- Ensure that the location of accesses is compatible with height of work stands and carts that will often be used.

Table 13.1 presents minimum access size requirements, expressed in inches, for one-handed tasks to be performed by a bare-handed maintenance person wearing regular clothes.[19]

TABLE 13.1

Minimum Access Size Requirements, Expressed in Inches, for One-Handed Tasks Performed by a Bare-Handed Maintenance Person Wearing Normal Clothes

Task Description	Dimensions in Inches	
	Height	Width
Placing arm through access up to the shoulder (i.e., full arm's length)	5	5
Placing arm through access up to the elbow	4	4.5
Inserting components/parts	1.75	4.5
Inserting a closed hand with thumb outside of fist	4.25	5.125
Inserting empty hand held flat	2.25	4.5

TABLE 13.2
Important Guidelines for Designing Modularized Systems or Products

Guideline Description

Divide the product/system/equipment under consideration into many modular units.

Make modules/parts/components as uniform in size and shape as feasible.

To the extent possible, design the modules for ease of operational testing when removed from the system or equipment.

Design all equipment so that an individual can easily replace any failed part.

Take an integrated approach to design by considering the problems of component design, modularization, and materials simultaneously.

Aim to make each module capable of being inspected independently and effectively.

Design each modular unit small and light enough so that a single individual can easily handle and carry it.

Ensure that the functional design of the equipment is matched with division of the equipment/product into removable and replaceable items.

Emphasize modularization for forward levels of maintenance to enhance operational capability.

Design control levers and linkages in such a way that allows easy disconnection from parts or components, in turn, simplifying component replacement process.

MODULARIZATION

Modularization may be described as the division of a product into functionally and physically distinct units to permit removal and replacement. The degree of modularization in a system or product depends on factors such as cost, practicality, and function. Every effort should be made to use modular construction wherever it is logistically feasible and practical as it helps reduce training costs, in addition to other concrete benefits. Table 13.2 presents some guidelines for designing modularized products.[19,20]

Some advantages of modularization are: relative ease in maintaining a divisible configuration, less time-consuming and -costly maintenance staff training, simplified new equipment design and shortened design time, easy to divide up maintenance responsibilities, lower skill levels and fewer tools required, existing product or equipment can be modified with the latest functional units replacing their older equivalents, fully automated approaches can be employed to manufacture standard "building blocks," and easy recognition, isolation, and replacement of faulty items leading to more efficient maintenance, thus lower equipment downtime.[19]

Disposable Modules

Disposable modules are designed to be discarded rather than repaired after a failure. They are used in situations when repair is costly or impractical. Their advantages outweigh the disadvantages, and maintainable modules require significant expenditure in materials, labor time, and tools.

The important benefits of a disposal-at-failure design include simpler and more concise trouble-shooting approaches; smaller, simpler, and more durable modules with a more reliable design; fewer types of spare parts required; reduction in required tools, personnel, facilities, and repair time; improved reliability due to the sealing and potting methods; and better standardization and interchangeability of modules.

Some of the drawbacks of a disposal-at-failure design are: an increase in inventory required because of need to have replacement modules on hand at all times, inability to redesign disposable modules, reduction in module performance and reliability because of production efforts to keep them inexpensive to justify their disposal, reduction in available data on maintenance and failures, and increase in unnecessary replacements.

INTERCHANGEABILITY

Interchangeability may be defined as an intentional aspect of design; any part/component/unit can be replaced within a given item by any similar part/component/unit. There are two distinct types of interchangeability: physical and functional. In physical interchangeability, two items can be connected, used, and mounted in the same location and in the same manner. With functional interchangeability, two given items serve the same function.

The basic principles of interchangeability include: liberal tolerances in the items requiring frequent replacement and servicing of parts because of wear or damage, that each part must be completely interchangeable with each other similar part, and that the items expected to function without part replacement strict interchangeability could be uneconomical.

The guidelines below are useful to achieving maximum interchangeability of parts and units in a given system:[2]

- Ensure that when physical interchangeability is a design characteristic, there is also functional interchangeability.
- When functional interchangeability is not desirable, there is no need for physical interchangeability.
- Avoid differences in size, shape, mounting, and other such characteristics.
- All parts/units/components expected to be identical should be completely interchangeable and identified as interchangeable.
- Provide sufficient information in job instructions and on plate identification for users to decide with a certain level of confidence if two like items are interchangeable.
- Provide adapters to make physical interchangeability possible in situations where complete (i.e., both functional and physical) interchangeability is not possible.
- Ensure that part/unit modifications do not change the method of mounting and connecting.

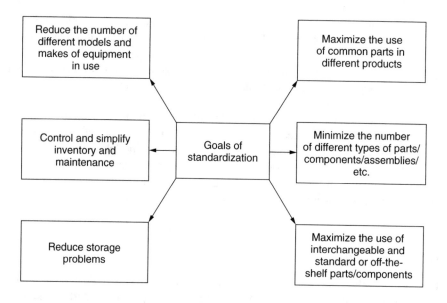

FIGURE 13.1 Main goals of standardization.

STANDARDIZATION

Standardization may be described as the attainment of maximum practical uniformity in an item's design.[19,21] Although standardization should be a central goal of design because use of nonstandard parts can result in lower reliability and increased maintenance, it must not be permitted to interfere with advances in technology or improvements in design. Important goals of standardization are presented in Fig. 13.1.

The advantages of standardization include:

- Reduction in design time, manufacturing cost, and maintenance time and cost
- Eliminates the need for special or close tolerance parts of components
- Useful to reduce errors in wiring and installation caused by variations in characteristics of similar items or units
- Useful to reduce the probability of accidents stemming from incorrect or unclear procedures
- Reduction in wrong use of parts or components
- Useful to facilitate "cannibalizing" maintenance approaches
- Improvement in reliability
- Reduction in procurement, stocking, and training problems

MAINTAINABILITY MEASURES AND FUNCTIONS

Various measures are used in maintainability analysis: for example, mean time to repair (MTTR), mean preventive maintenance time, and mean maintenance downtime. Maintainability functions are used to predict the probability that a repair, starting at time

$t = 0$, will be completed in a time t. Some maintainability measures and functions are presented below.[2,5,22–24]

MEAN TIME TO REPAIR

Mean time to repair (MTTR) is probably the most widely used maintainability measure. It measures the elapsed time required to perform a given maintenance activity. MTTR is expressed by

$$\text{MTTR} = \left(\sum_{i=1}^{k} \lambda_i \text{CMT}_i \right) \Big/ \sum_{i=1}^{k} \lambda_i \qquad (13.1)$$

where

k = number of units or parts,
λ_i = failure rate of unit/part i, for $i = 1, 2, 3,\ldots,k$,
CMT_i = corrective maintenance/repair time required to repair unit/part i, for $i = 1, 2, 3,\ldots,k$.

Usually, times to repair follow exponential, lognormal, and normal probability distributions.

Example 13.1

A piece of electronic equipment is composed of five replaceable subsystems 1–5, with corresponding failure rates: $\lambda_1 = 0.0004$ failures per hour, $\lambda_2 = 0.0005$ failures per hour, $\lambda_3 = 0.0006$ failures per hour, $\lambda_4 = 0.0007$ failures per hour, and $\lambda_5 = 0.0008$ failures per hour, respectively. The corresponding corrective maintenance times for subsystems 1–5 are $T_1 = 2$ h, $T_2 = 3$ h, $T_3 = 4$ h, $T_4 = 5$ h, and $T_5 = 6$ h, respectively. Calculate the equipment MTTR.
By substituting the given data into Eq. (13.1), we get

$$\text{MTTR} = \frac{(0.0004 \times 2) + (0.0005 \times 3) + (0.0006 \times 4) + (0.0007 \times 5) + (0.0008 \times 6)}{0.0004 + 0.0005 + 0.0006 + 0.0007 + 0.0008}$$

$$= 4.33 \text{ h}$$

The electronic equipment mean time to repair is 4.33 h.

MEAN PREVENTIVE MAINTENANCE TIME

To keep equipment at a specified performance level, performance of preventive maintenance activities such as inspections, tuning, and calibrations are essential. Usually, a well-planned preventive maintenance program plays an instrumental role in reducing equipment downtime and improving its performance.
The mean preventive maintenance time is defined by

$$\text{MPMT} = \left(\sum_{i=1}^{m} \text{FPM}_i \times \text{ETPMT}_i \right) \Big/ \sum_{i=1}^{m} \text{FPM}_i \qquad (13.2)$$

where

MPMT = mean preventive maintenance time,
m = total number of preventive maintenance tasks,
FPM_i = frequency of preventive maintenance task i, for $i = 1, 2, 3,...,m$,
$ETPMT_i$ = elapsed time for preventive maintenance task i, for $i = 1, 2, 3,...,m$.

In Eq. (13.2), note that if the frequencies FPM_i are given in maintenance tasks per hour, then $ETPMT_i$ should also be given in hours.

MEAN MAINTENANCE DOWNTIME

Mean maintenance downtime (MMD) may be described as the total time required either to restore system to a given performance level or to keep it at that level of performance. It is composed of corrective maintenance, preventive maintenance, administrative delay, and logistic delay times.

The administrative delay time is the system/item downtime due to administrative constraints. Logistic delay time is the time spent waiting for a required resource such as a spare part, a specific test, or a facility.

MMD is defined by

$$MMD = MAMT + LDT + ADT \tag{13.3}$$

where

ADT = administrative delay time,
LDT = logistic delay time,
MAMT = mean active maintenance time or mean time needed to perform preventive and corrective maintenance-associated tasks.

MAINTAINABILITY FUNCTIONS

Maintainability functions predict the probability that a repair, starting at time $t = 0$, will be completed in a time t. The maintainability function for any distribution is defined by

$$M(t) = \int_0^t f_R(t)\, dt \tag{13.4}$$

where

t = time,
$M(t)$ = maintainability function,
$f_R(t)$ = probability density function of the repair time.

Maintainability Function: Exponential Distribution

Exponential distribution is widely used in maintainability work to represent repair times. Its probability density function is expressed by

$$f_R(t) = \left(\frac{1}{MTTR}\right)\exp\left(-\frac{t}{MTTR}\right) \tag{13.5}$$

Inserting Eq. (13.5) into Eq. (13.4), we obtain

$$M(t) = \int_0^t \left(\frac{1}{\text{MTTR}}\right) \exp\left(-\frac{t}{\text{MTTR}}\right) dt$$

$$= 1 - \exp\left(-\frac{t}{\text{MTTR}}\right) \tag{13.6}$$

Equation (13.6) is maintainability function for exponentially distributed times to repair.

Example 13.2

Assume that MTTR of an electronic system is 4 h. Determine the probability that a repair action will be accomplished in 8 h, if the repair times are exponentially distributed.

Substituting the given values into Eq. (13.6) yields

$$M(6) = 1 - \exp\left(-\frac{8}{4}\right) = 0.8647$$

There is an approximately 87% chance that the repair will be accomplished in 8 h.

Maintainability Function: Weibull Distribution

Weibull distribution can be used to represent times to repair. Its probability density function is defined by

$$f_R(t) = (\beta/\theta^\beta)t^{\beta-1} \exp[-(t/\theta)^\beta] \tag{13.7}$$

where
β = shape parameter,
θ = scale parameter.

Substituting Eq. (13.7) into Eq. (13.4), we get

$$M(t) = \int_0^t (\beta/\theta^\beta)t^{\beta-1} \exp[-(t/\theta)^\beta] \, dt$$

$$= 1 - \exp[-(t/\theta)^\beta] \tag{13.8}$$

For $\beta = 1$ and $\theta = $ MTTR, Eq. (13.8) reduces to Eq. (13.6). For $\beta = 2$, Eq. (13.8) is the maintainability function for Rayleigh distribution.

Maintainability Function: Gamma Distribution

Gamma distribution is sometimes used to represent various types of maintenance time data. The distribution probability density function is defined by

$$f_R(t) = \frac{\lambda^\beta}{\Gamma(\beta)} t^{\beta-1} e^{-\lambda t} \tag{13.9}$$

where
λ = scale parameter,
β = shape parameter,
$\Gamma(\beta)$ = gamma function and is expressed by

$$\Gamma(\beta) = \int_0^\infty y^{\beta-1} e^{-y}\, dy \tag{13.10}$$

Substituting Eq. (13.9) into Eq. (13.4), we get

$$M(t) = \frac{\lambda^\beta}{\Gamma(\beta)} \int_0^t t^{\beta-1} e^{-\lambda t}\, dt \tag{13.11}$$

Since $\Gamma(1) = 1$, at $\beta = 1$, Eq. (13.11) becomes the maintainability function for the exponential distribution.

Maintainability Function: Erlangian Distribution

The Erlangian distribution is the special case of the gamma distribution when the gamma distribution shape parameter takes positive integer values. Thus, from Eq. (13.10) we get

$$\Gamma(\beta) = (\beta - 1)! \tag{13.12}$$

From Eq. (13.9), we write the following probability density function for the Erlangian distribution:

$$f_R(t) = \frac{\lambda^\beta}{(\beta-1)!} t^{\beta-1} e^{-\lambda t} \tag{13.13}$$

Inserting Eq. (13.13) into Eq. (13.4) yields

$$M(t) = \int_0^t \frac{\lambda^\beta}{(\beta-1)!} t^{\beta-1} e^{-\lambda t}\, dt$$

$$= 1 - \sum_{i=0}^{\beta-1} [e^{-\lambda t}(\lambda t)^i / i!] \tag{13.14}$$

Maintainability Function: Normal Distribution

Normal distribution can be used to represent times to repair. Its probability density function is given by

$$f_R(t) = \frac{1}{\sigma\sqrt{2\pi}} \exp\left[-\frac{1}{2}\left(\frac{t-\theta}{\sigma}\right)^2\right] \tag{13.15}$$

where
σ = standard deviation of the variable maintenance time t around the mean value θ,
θ = mean of maintenance times.

Substituting Eq. (13.15) into Eq. (13.4), we obtain

$$M(t) = \frac{1}{\sigma\sqrt{2\pi}} \int_0^t \exp\left[-\frac{1}{2}\left(\frac{t-\theta}{\sigma}\right)^2\right] dt \qquad (13.16)$$

The mean of the maintenance times is expressed by

$$\theta = \sum_{i=1}^{k} t_i/k \qquad (13.17)$$

where
k = total number of maintenance tasks performed,
t_i = ith maintenance time, for $i = 1, 2, 3, ..., k$.

The standard deviation is given by

$$\sigma = \left[\sum_{i=1}^{k} (t_i - \theta)^2/(k-1)\right]^{1/2} \qquad (13.18)$$

COMMON ERRORS RELATED
TO MAINTAINABILITY DESIGN

Often equipment designers make various design errors that adversely affect system maintainability. Some of those design errors are listed below.[25]

- Adjustments placed out of reach of maintenance personnel and existence of inadequate room for workers to make appropriate adjustments when wearing a glove
- Access doors contain too many small screws and wrong or no handles
- Difficult to locate adjusting screws
- Low-reliability test equipment falsely reports product failures
- Adjusting screws too close to an exposed power supply terminal or a hot part
- Low-reliability parts installed beneath other parts, forcing maintenance individuals to disassemble other parts to reach them
- Fragile parts placed just within the lower edge of the chassis, making them more likely to be broken accidently
- Screwdriver-oriented adjustments placed beneath modules in such a manner that repair personnel find it difficult to reach them

- Removable parts installed in such a way that repair personnel find it impossible to remove them without dismantling the entire unit from its case or removing other items
- Different modules designed with identical sockets and connectors, thus increasing the risk of installing modules in the wrong place
- Subassemblies screwed together in such a manner that maintenance personnel find it impossible to distinguish what is being held by each screw

PROBLEMS

1. Write an essay on historical developments in maintainability.
2. Define the following terms:
   - Maintainability
   - Maintainability function
   - Downtime
3. Discuss maintainability management in the system life cycle.
4. List at least ten maintainability design characteristics.
5. Describe in detail the following:
   - Accessibility
   - Modularization
   - Interchangeability
6. List the primary goals of standardization
7. A system is composed of three replaceable subsystems 1–3, with failure rates: $\lambda_1 = 0.0001$ failures per hour, $\lambda_2 = 0.0002$ failures per hour, and $\lambda_3 = 0.0003$ failures per hour, respectively. The respective corresponding corrective maintenance times for subsystems 1–3 are $T_1 = 1$ h, $T_2 = 2$ h, and $T_3 = 3$ h, respectively. Calculate the system mean time to repair.
8. Obtain maintainability functions for the following times to repair probability density functions:
   - Exponential
   - Lognormal
   - Gamma
9. List at least ten common maintainability design errors.
10. Write the Erlangian probability density function of times to repair.

REFERENCES

1. Smith, D.J. and Babb, A.H., *Maintainability Engineering,* John Wiley & Sons, New York, 1973.
2. AMCP-706-133, *Engineering Design Handbook: Maintainability Engineering Theory and Practice,* Department of Defense, Washington, D.C., 1976.
3. Retterer, B.L. and Kowalski, R.A., Maintainability: a historical perspective, *IEEE Transac. Reliability,* 33, 1984, 56–61.
4. SAE G-11, *Reliability, Maintainability, and Supportability Guidebook,* The Society of Automotive Engineers, Warrendale, Pennsylvania, 1990.

5. Dhillon, B.S., *Engineering Maintainability,* Gulf Publishing Co., Houston, Texas, 1999.
6. Ankenbrandt, F.L., ed., *Electronic Maintainability,* Engineering Publishers, Elizabeth, New Jersey, 1960.
7. MIL-HDBK-472, *Maintainability Prediction,* Department of Defense, Washington, D.C., 1966.
8. MIL-STD-470, *Maintainability Program Requirements,* Department of Defense, Washington, D.C., 1966.
9. MIL-STD-471, *Maintainability Demonstration,* Department of Defense, Washington, D.C., 1966.
10. Dhillon, B.S., *Reliability and Quality Control: Bibliography on General and Specialized Areas,* Beta Publishers, Gloucester, Ontario, Canada, 1993.
11. Dorf, R.C., ed., *Technology Management Handbook,* CRC Press, Boca Raton, Florida, 1999.
12. Niebel, B.W., *Engineering Maintenance Management,* Marcel Dekker, New York, 1994.
13. MIL-STD-721 C, *Definitions of Terms for Reliability and Maintainability,* Department of Defense, Washington, D.C.
14. Omdahl, T.P., ed., *Reliability, Availability and Maintainability (RAM) Dictionary,* ASQC Quality Press, Milwaukee, Wisconsin, 1988.
15. Von Alven, W.H., ed., *Reliability Engineering,* Prentice-Hall, Englewood Cliffs, New Jersey, 1964.
16. Naresky, J.J., Reliability definitions, *IEEE Transac. on Reliability,* 19, 1970, 198–200.
17. Dorf, R.C., ed., *Technology Management Handbook,* CRC Press, Boca Raton, Florida, 1999.
18. Niebel, B.W., *Engineering Maintenance Management,* Marcel Dekker, New York, 1994.
19. AMCP 706-134, *Engineering Design Handbook: Maintainability Guide for Design,* Department of Defense, Washington, D.C., 1972.
20. Altman, J.W., et al., Guide to Design of Mechanical Equipment for Maintainability, Report No. ASD-TR-61-381, U.S. Air Force Systems Command, Wright-Patterson Air Force Base, Ohio, 1961.
21. Ankenbrandt, F.L., et al., *Maintainability Design,* Engineering Publishers, Elizabeth, New Jersey, 1963.
22. Grant-Ireson, W. and Coombs, C.F., eds., *Handbook of Reliability Engineering and Management,* McGraw-Hill, New York, 1988.
23. Blanchard, B.S., Verma, D., and Peterson, E.L., *Maintainability,* John Wiley & Sons, New York, 1995.
24. Blanchard, B.S., *Logistics Engineering and Management,* Prentice-Hall, Englewood Cliffs, New Jersey, 1981.
25. Pecht, M., ed., *Product Reliability, Maintainability, Supportability Handbook,* CRC Press, Boca Raton, Florida, 1995.

Index